ABOUT ADDICTIONS:

NOTES FROM PSYCHOLOGY,

NEUROSCIENCE AND NLP

[REVISED]

RICHARD M. GRAY, PH.D.

Copyright 2014

Richard M. Gray, Ph.D.

All Rights Reserved

Library of Congress-cataloging in-publication-data

Gray, Richard M.

About Addictions: Notes from Psychology, Neuroscience and NLP.

Richard Gray.—Revised edition.

Includes bibliographical references.

ISBN 978-1-312-20755-4

Addiction and recovery. 2. Psychology. 3. Neuro-Linguistic Programming.

TABLE OF CONTENTS

Introduction	1
Organization of the text	5
A Brief Introduction to NLP	9
Presuppositions	11
Chapter Two: The Power of the Name:	25
Diagnosis—addiction spectrum disorders.	25
Phenotypic variation	35
Further complications	39
Chapter Three: Three Important Studies	45
Study number one, rat parks.	47
Study Number Two, Vietnam Veterans	51
Study Number Three Pain, Opioids and Addiction	57

Chapter Four: Mechanisms of Motivation and Reward:	61
The addictive properties of drugs of abuse	62
The action of drugs on basic neural function	62
Addiction, Dependence and the midbrain dopamine system	66
Defects in Inhibition and Behavioral Monitoring	73
Dissociation of behavioural networks	75
Target Problems	76
Chapter Five: Dimensions of Motivation: A motivational primer	77
Motivations	80
Motivation and Addiction	91
Chapter Six: Hierarchies and preferences	93
Dilts' Neurological Levels	104
Chapter Seven: Stages of Change and MET	115
Chapter Eight: Outcomes	125
Levels of Motivation	132
Assessing and Creating Well-formed outcomes	136
Tools for creating intrinsically meaningful outcomes	137
Asking	137
End state energy	139

Outcome sequiturs	143
Chapter Nine: NLP techniques for motivated clients	**147**
Asking	148
Permissions	150
Creating anchored resource states	153
Chapter Ten: Pseudo-orientations in time	**169**
Working through a well-formed outcome	169
The miracle question	187
Chapter Eleven: The submodality blowout	**193**
Outline	**197**
Chapter Twelve: Changing the unwilling: The Brooklyn Program	201
Chapter Thirteen: Postscript	211
The NLP Approach	217
References	227

Introduction

This is a short book about drugs and drug treatment culled from more than 25 years of working in the criminal justice system. During the last ten years of that period, I worked exclusively with addicts, users and abusers. Throughout the last eight years, I developed and ran a program based on the principles of Neuro-Linguistic Programming (NLP) for the treatment of substance use disorders. The program was in operation until my retirement in 2004, receiving both national and international recognition. During the ensuing ten years, I have attempted to keep up with the relevant brain science and have coached many providers in the use of the program.

During the entire 27 years of my tenure in law enforcement, I had regular experience with persons who used or were addicted to illegal and legal mind-altering substances. During most of my career, I trusted the experts. As time when on, however, I began to notice a pattern that disturbed me greatly. It seemed that no matter where I turned, with the exception of a few religious programs—and a few *very special* service providers, everyone seemed to have the same answers for what appeared to me to be very different levels of problems. No matter what the substance, or the level of use, everyone was labeled as an addict and sent to detox, then to rehab or long term inpatient care, followed by intensive outpatient treatment, complimented by interminable 12- step meetings. At first, the inpatient treatment consisted largely of humiliation, education and 12-step work. As time went on, it moved more towards counseling, education and 12-step work. Everyone got the same thing.

As all of this went on, there seemed to be a mantra thrumming in the background: use always becomes abuse, abuse always becomes addiction, and addiction always ends in abstinence or death. There seemed to be no appreciation for the subtleties of diagnosis and no possibility of recreational use. Because most of the substances were illegal, any use qualified the user as an abuser. If they continued to use for more than six months and because they did so in violation of the law and so endangered their livelihood and relationships[1] they met the diagnostic criteria for substance abuse disorder.

In spite of this, it was plain to me that many of my clients were casual users, or regular but non-problematic users. The answer of the industry was that they were in denial and I was probably an enabler. It seemed as though I was regularly met by the now famous challenge: Who are you going to believe, me or your lying eyes?

In 1993, I was reassigned from a specialist position as Automation Liaison, to a drug treatment caseload. I was suddenly confronted with a rotating group of between 30 and 50 persons under

---

[1] The fact is that only ten to twelve percent of people who try drugs get hooked and of those about eighty percent stop using on their own.

federal supervision who had histories of *addiction spectrum disorders*[2], or who were suspected of having them. Since I knew little more about addictions than the standard doctrine related above, I immersed myself in scientific journals and sought to discover the most scientifically accurate information available. What I found was that it was all wrong.

At this point, I had the positive advantage of training in Neuro-Linguistic Programming (NLP) and began to construct my own approach. This approach was based upon the best scientific evidence that I could find and tools from NLP and Ericksonian hypnosis. From Prochaska's Stages of Change Model, I learned that the most important element in recovery was having a meaningful future that was strong enough to compete with the problem behavior. From the work of Richard Bandler came the idea that a present time positive resource that was stronger, more enjoyable and more intuitively accessible could reframe addictions into irrelevance. From the work of Milton Erickson, Carl Jung and Abraham Maslow, I learned that whole lives could be reframed by just the kinds of futures and strong positive affects that Bandler and Prochaska had suggested. From Connirae and Tamara Andreas came the idea that powerful experiences having a spiritual impact could be structured for anyone.

From this mix, the basic presuppositions of NLP, a significant helping of neurophysiology and a belief in the fundamental wholeness of all people, I created the Brooklyn Program. The program boasted 26.9% abstinence rates for verified users one year after completion and this came at the cost of only two facilitator hours per week and a

---

[2] *Addiction spectrum disorders* is my term for the various layers of substance use disorders. It extends to behavioral addictions and compulsions like gambling and sexual addictions and is somewhat more elegant than the standard terminology. It will be used throughout the book. As of this revision it matches with the restructuring of the addiction diagnoses in DSM 5.

savings of about $3000 per successful participant. The program received both national and international attention. This book, however, is not about that program. Although we will review its basic principles, this book is focused more broadly[3].

The following material is designed to provide the reader with tools for thinking about addictions and a small number of tools from the NLP toolbox that I have used and that have been used by others to work with addiction spectrum disorders. It is not an exhaustive exposition on drugs or addiction, or even of the possible treatment modalities. It is designed to provide tools for thought so that an informed set of professional level distinctions can be made about the problems encountered and some of the things that can be done for them. All of the interventions suggested have worked for various people in various contexts; none of them has worked for everyone. There are no panaceas.

With the notable exception of the Stages of Change Model and Motivational Interviewing, this book does not treat any of the standard approaches to addiction. The 12-step models—whatever their value— have been reviewed sufficiently elsewhere. Cue extinction models, community reinforcement and other behavioral models are also not treated. Radical detoxification and pharmacological treatments are not covered. Most of the tools here are rooted in NLP, a field that has much to offer and that has received far too little attention from the professional world of addictions treatment.

In this new edition I have updated the information from neuroscience and corrected some of the egregious editing failures in the original.

---

[3] Interested readers can download a free copy of *Transforming Futures: The Brooklyn Programs Facilitators Manual* (2nd ed.) from: http://www.lulu.com/shop/search.ep?contributorId=424734

ORGANIZATION OF THE TEXT

Chapter One: A Brief Introduction to NLP, is a brief introduction to NLP with an emphasis on its early history and its basic presuppositions. It is not intended to be complete in any sense; complete introductory texts are referenced in the references.

Chapter Two: The Power of the Name, discusses diagnosis, misdiagnosis and a practical means of sorting through the issues related to defining the client's problem in a meaningful way. While NLP is generally not diagnosis driven, a familiarity with the language of the addictions profession is crucial as we begin to seek wider recognition in the field. In the new edition I have added some neurological problems that may contribute to the diagnostic issues.

Chapter Three: Three Important Studies, deals with three studies of 'addictive substances' that most people will find surprising. These studies have often been ignored by addictions treatment providers and policy makers. They are, nevertheless, crucial for forming a complete understanding of the relationship between drugs and addiction to drugs.

Chapter Four: Mechanisms of Motivation and Reward, is a basic overview of the neurophysiology of addiction and motivation. The information here is up-to-date and has been crucial in transforming my own idea of what addiction spectrum disorders are about. This information provides deep validation for many of the ideas that are central to NLP's understanding of human nature.

Chapter Five: Dimensions of Motivation: A motivational primer, discusses some very basic ideas about how people are motivated and what works to motivate people. It also differentiates between relatively superficial motivations and deeper transformative motivations. It is a distinction that NLPers will find instructive and useful

Chapter Six: Hierarchies and preferences, picks up the idea of preference hierarchies from the materials on neurophysiology and motivation and discusses how preferences, values and motivations are ordered. It makes use of, and clarifies, Robert Dilts' neurological levels to understand the cascade of influences as motivations flow from one level to another.

Chapter Seven: The Stages of Change and MET, is a discussion of the Stages of Change Model, the strong principle of change and Motivational Interviewing or Motivational Enhancement Therapy (MET). Although these are not NLP focused, they are some of the best-validated approaches to understanding change and treatment in use today. As such, they represent a means of understanding addiction and change that are professionally important. The treatment of MET consists of a brief outline of the practice. It is presented only to familiarize practitioners with a treatment approach that they will be encountering while noting that it is a place where NLP stands to make a significant contribution.

Chapter Eight: Outcomes, discusses outcomes in depth. Although the well-formedness conditions for outcomes has been briefly discussed in previous sections, it is given more attention here especially with regard to its violation in most treatment contexts. Attention is also given to techniques for helping people to create intrinsic, well-formed outcomes. These include asking, finding and using end state energy and a brief description of Connirae and Tamara Andreas' Core Transformation technique with modifications.

Chapter Nine: NLP techniques for motivated clients, begins a discussion of NLP treatments keyed to the non-standard diagnostic criteria developed in the first chapter. The focus here is on people who want help and for whom meta-model challenges, permissions, and compulsion blowouts are appropriate.

Chapter Ten: Pseudo-orientations in time, treats one of the signal contributions of Milton Erickson and its application to addictions.

Chapter Eleven: The submodality blowout, provides a brief summary of the technique as provided by Steve Andreas, one of the founding lights in NLP, and attempts to tease out some guiding principles for facilitating change with addiction spectrum disorders.

Chapter Twelve: The Brooklyn Program, is a brief summary of the author's award-winning program for mandated treatment populations

Chapter Thirteen: Postscript, provides a brief summary of the materials presented in hope of providing a comprehensive and integrative frame for thinking about addictions. The second half of the chapter provides an outline of the NLP approach, independent of techniques.

May, 2014

Long Branch, NJ

Chapter 1

A Brief Introduction to NLP

Neuro-Linguistic Programming (NLP) is a set of tools comprising an epistemology, a methodology and a set of techniques rooted in a strategy for modeling human behavior, developed in the mid 1970's by linguist John Grinder and Psychology students Richard Bandler and Frank Pucelik (Grinder & Pucelik, 2012; Thomas Yeager, Personal Communication, 2007). Beginning by modeling the therapeutic styles of Frederick Perls, Bandler and Pucelik soon found that they could analyze those therapeutic insights into teachable patterns. Grinder, then assistant professor of Linguistics at the University of California, Santa Cruz, was invited to join the groups started by Pucelik and Bandler in order to help with the systemization of these findings.

In his adaptation of transformational grammar, Grinder understood that the structure of both language and experience could be modeled in terms of sequences of sensory experience including what was seen, heard, felt, smelled or tasted: the Visual, Auditory, Kinesthetic, Olfactory and Gustatory (VAKOG) elements. When accurately mapped, these sequences would provide the keys not only to modeling the subject behavior but also to modifying unwanted or non-useful behaviors (Bandler and Grinder 1975, 1979; Bostic St. Clair & Grinder, 2002; Dilts, 1985; Dilts, Bandler et al., 1980; Dilts & Delozier, 2000).

Bandler was described by Grinder as natural therapist who had the unique skill of being able to learn and quickly master almost any psychotherapeutic technique. As their collaboration with Pucelik and others began, Bandler would experientially master a psychotherapeutic approach and together they would parse the more salient aspects of the techniques involved in terms of Grinder's model (Bostic St. Clair & Grinder, 2002).

As he began attending the groups, Grinder found regular use of verbal patterns already well known to linguists and other patterns, which he documented as techniques for behavioral change. Over the next several years, often at the urging of Gregory Bateson, Grinder and Bandler applied their modeling skills to the patterns and techniques of Virginia Satir, founder of Conjoint Family Therapy and founding member of the Mental Research Institute in Palo Alto, California; Milton Erickson, often described as the father of modern hypnotherapy, and others. In the course of their researches, they created a technique for modeling behavior and a series of tools of general applicability in addition to interventions for specific pathologies, learning problems and behavioral issues. This basic repertoire was enhanced significantly by the contributions of other early participants in the development of NLP, including Robert Dilts, another of Grinder's graduate students; John and Connirae Stevens (Steve and Connirae Andreas), already well known in Gestalt circles; Leslie (Cameron-Bandler) Lebeau, Judith Delozier; David Gordon and Steven Gilligan (Bandler and Grinder 1975, 1979; Bostic St. Clair & Grinder, 2002; Dilts, Bandler et al., 1980; Dilts, Delozier & Delozier, 2000; Grinder & Pucelik, 2012; Lewis and Pucelik, 1990; O'Connor and Seymour, 1990).

NLP can be understood as an approach to modeling excellence and a way of describing the models so that they become replicable and testable; a set of practical techniques for dealing with specific problems and issues, and a set of basic skills and techniques that may be thought of as the basic NLP tool kit. In general, this tool-kit represents the elements of most NLP-based interventions. These include accessing cues, sensory-based predicates, the meta-model, the milton-model, pacing and leading, anchoring, reframing, change personal history, Visual-Kinesthetic Dissociation and state management. For our purposes, submodalities are considered essential elements of the tool kit (Gray, 2008a).

PRESUPPOSITIONS

NLP is characterized by a set of presuppositions that outline its unique approach to communication and change. Presuppositions are things that one takes for granted. They are the givens that inform the way one interacts with the world. According to Judith Delozier, one of the founders of the field, the presuppositions are the heart of NLP. If you take them seriously, the world becomes a very different place (Bandler and Grinder 1975, 1979; Bostic St. Clair & Grinder, 2002; Dilts, 1985; Dilts, Bandler et al., 1980; Dilts, Delozier & Delozier,

2000; IASH & Delozier, 2006). Steve Andreas has proposed the following list as a consensus model:

1. People respond to their map of reality, not to reality itself. 2. Every behavior is motivated by positive intent. 3. People always make the best choice available to them at the time. 4. Every behavior is useful in some context. 5. Communication is expressed in all three major sensory modalities in addition to words. 6. The meaning of your communication is indicated by the response that you get. 7. People function systematically, according to a structure. 8. Choice is better than no choice. Always add choices; never try to subtract them. 9. Anyone can learn to do anything that someone else can do. 10. People already have all the resources they need. 11. There is no such thing as failure, only feedback. 12. If we break a task down into small enough chunks, any task can be accomplished. 13. Satisfying objections is necessary for lasting change. Respect "resistance" (Andreas, 2006).

These presuppositions are taken as axiomatic in the NLP world and find their origins in various perspectives.

People respond to their map of reality, not to reality itself: the map is not the territory

The phrase itself comes from the work of Alfred Korzybski who was the founding light in the field of General Semantics. It has been suggested that the source was actually Gregory Bateson (Tosey & Mathison, 2009). In his thought, we speak in terms of very personal perceptions and should be very careful in what we take to be objective knowledge. In the world of professional action, it is crucial that we carefully test to discover what words really mean for the person who uses them.

People are in the habit of thinking that words and labels are identical with the things they denote. We argue over words and nuances of meanings. We kill people because their understanding of a word is not the same as ours. On a practical level it means that we can never assume that what someone means when they use a word is the same as our own understanding. People raised in America and Western Europe can distinguish several million shades of color. *Colr.org* lists 22,607 named shades. So, the question naturally arises, when I say green, to which of the 949 shades of green listed there (Colr.org) do I refer?

On another level, we are constantly responding to the world of virtual entertainment as if it were real. My children used to get great enjoyment over watching me jump whenever the spiders appeared in a movie from their childhood. It was not uncommon to find me calling out to the people on the screen that there was something around the corner and wincing in pain at the televised blows of a fist-fight. The map is not the territory. How many Sunday afternoon quarterbacks are thrown into a frenzy over the televised shenanigans of their favorite football game? Riots in soccer stadia attest to the loss of proper perspective.

Richard Bandler has suggested, following Bateson, that when we mistake the map for the territory we might as well sit down in a restaurant and eat the menu. The map is not the territory (Bandler & Grinder, 1975, 1979; Bateson, 1972).

Every behavior has a positive intent.

This presupposition suggests both the perspectives of reinforcement theory and evolutionary psychology. Even though a behavior may be incomprehensible to an external observer, it has meaning within the life context of the individual. Ultimately it is purposeful in terms of fulfilling some need or protecting from some possible injury. It is immaterial whether those needs or cares are real or imagined; insofar as they are real to the individual, they provide a context for understanding their behavior and a basis for designing change.

This presupposition is strongly supported by evolutionary psychology and social anthropology. It is generally understood, that many of our behavioral patterns, especially those rooted in the perception of emotion and value evolved for very different times than our own. These often provide value to behaviors that make no sense to most of us who are now living in an urbanized, mechanized, electronically connected world. They nevertheless make sense to the person experiencing them. From the perspective of embodied consciousness, these archaic patterns lie at the heart of our metaphorical understanding of the world. (Andreas, 2006; Goodwyn, 2012; Haule, 2010; Lakoff & Johnson, 1980).

Every purposeful behavior moves towards some desired outcome and makes sense to the actor, even if it makes no sense to victims, witnesses, therapists, and law enforcement officers. When we understand how an act makes sense to the actor, we gain valuable insight into why they did what they did and clues about how to keep it from happening again.

People always make the best choice available to them at the time.

Options are ordered in terms of their incentive salience. That is, faced with a set of choices, every context defines one or another option as more important, more relevant, or desirable than the others.

This is a function of context, reinforcement history and the perceptual structure of the stimuli signifying the option. History, the state of the organism and context affect the internal structure of experience (submodalities) and determine the relative value of the presented options. Other people's choices can only be understood in terms of their maps. To us, as outsiders, they will make no sense until we have comprehended their map of the territory with their meanings and preferences. We cannot do this without asking, and listening, and checking with the client to ensure that we've gotten it right.

Every behavior is useful in some context.

The utility of a behavior is often defined by its contextual frame. It is important here to distinguish the behavior from its target. Violence is plainly useful in self-protection. When framed as part of the successful execution of a crime it is useful to the offender. The neurotic checking and rechecking of switches and controls is symptomatic of OCD but crucial for preflight checks. Determining where a behavior is useful, reframing, is an important therapeutic skill. Erickson called this the principle of utilization.

Communication is expressed in all three major sensory modalities in addition to words.

We often think of communication in terms of language and intentional signifiers, however, there exists a whole range of paralinguistic signals that do not merely modulate the meaning of linguistic communication but constitute separate channels in themselves. Tonality, pace, punctuation and word sequence define some of the range of non-verbal communication in the auditory sphere. Posture, facial expressions—including micro expressions, gestures and physiological indicators (pulse, perspiration, tremors), provide kinesthetic and visual communications that can be crucial (Andreas, 2006; Andreas, C. & Andreas, S. ,1989; Andreas, S. & Andreas, C. ,1987; Bandler & Grinder, 1975a, b, 1979; Ekman & Frank, 1993; Ekman & Friesen, 1972; Vrij & Mann, 2004). Relative to this general principle, Paul Watzlawick said that you cannot not communicate (Watzlawick, Beavin & Jackson, 1967). Mehrabian (1972) showed that in emotional communication, more than 90 percent of the emotional content is communicated by non-linguistic channels.

The meaning of your communication is the response that you get.

NLP takes the radical position that you are responsible for the outcomes of your communications. If someone misunderstands you, you must have or develop the flexibility to change your communication so that your message gets through. Animal trainers, behaviorists and performers have known this for a long time. If my act does not get a standing ovation, I must be doing something wrong. If the dog won't learn the trick, it MUST be my fault, not the dog's.

We have all been in the position where we have complained that someone should know what I mean and is just being perverse by not understanding, or that they are just being perverse in their misinterpretation of my words. In NLP, we cannot make this claim. The fault of miscommunication is never in my listener, the responsibility for successful communication always falls to me.

NLP focuses on the pragmatics of communication. It requires the communicator to have an outcome or purpose for the communication, success criteria for knowing if that outcome has been achieved and sufficient flexibility to do something different if what they've been doing doesn't work. Communication is an evidence-based procedure. It requires multiple levels of awareness and a clear understanding of where you want to go. If communication it is not going where you want it to go, it is up to you to do something different. Your listener will tell you how well you are doing by their response to your words and actions.

People function systematically, according to a structure.

NLP assumes that behaviors are related to one another in systematic ways. These are sometimes termed strategies. In mainline psychology they are identical to schemas and scripts. In terms of basic NLP concepts, scripts and schemas are larger examples of the ordering principle revealed as syntax and well-formedness conditions. There is a tendency for behaviors at all levels of integration to streamline along habitual patterns and sequences. This is revealed in the behavioral phenomena of chaining individual behaviors and responses into complex sequences and the function of the basal ganglia as the neurological center that assembles typical behaviors. This presupposition is closely related to the first--People respond to their map of reality, not to reality itself--in that behavior shapes perception and expectation and is therefore a major determinant of how each individual views the world. NLP is implicitly systems theoretical, behaviors and perceptions are not lumps or simple aggregates of components, but organized systems in which the interrelation of subsystems gives rise to properties that could not be predicted by a simple inventory of the parts. Lives are systems and Robert Dilts' Hierarchy gives expression to this insight as do Bateson's levels of learning (Andreas, 2006; Bateson, 1972; Bertalanffy, 1968; Dilts, 1994-5; Fidler, 1982; Gray, 1996; Gray, 2010; Gray, 2011; Piaget, 1970).

Choice is better than no choice.

Early on, the founders adopted the law of requisite variety from cybernetics and evolutionary biology (Ashby, 1956; Bandler & Grinder, 1975, 1979). The principle states that the organism with the most options in a given environment is the most likely to survive. This extends to humans in the observation that pathology often exists in terms of limited options or what NLP calls stuck states. The correlate of this presupposition is if what you are doing doesn't work, do something different. The presupposition, however, does not suggest that the unlimited multiplication of options is a good thing in itself—option paralysis (Iyengar & Lepper, 2000) is a real phenomenon. Options should be multiplied so as to facilitate flexibility in an appropriate manner.

Anyone can learn to do anything that someone else can do.

At the essential core of NLP are the paired ideas that every behavior can be analyzed into a sequence of sensory elements (VAKOG) and that those sequences constitute the essence of the behavior, it then follows that it is possible to model any behavior and teach it to anyone else. This is modeling, the heart of the NLP enterprise. The vast majority of the techniques and practices that are commonly identified as the core NLP practices are derived from models of behaviors exhibited by exemplars like Milton Erickson, Fritz Perls and Virginia Satir (Andreas, C. & Andreas, S., 1989; Andreas, S. & Andreas, C., 1987; Bandler & Grinder, 1975, 1979).

People already have all the resources they need.

There is a presupposition in normal experience that some people either have or do not have certain abilities and capacities. In general, NLP holds that every capacity can be analyzed into structural units or chunks which, when arranged according to the correct syntactic relations will allow the individual to accomplish virtually anything that is available to any other similarly endowed human. Every psychologically whole individual has all of the resources that they need to do whatever they can imagine. Because NLP is a modeling discipline, people can use the behaviors of others as models for their own behavior. Given the fact that imagined and observed experience constitutes some level of practice, imaginal and in vivo role play makes any observed behavior a real resource. The examples of Helen Keller, George Washington Carver, and the blind and limbless mountain climbers of recent years are potent examples. The author who overcame her autism by learning to make lists that allowed her to move through the social world is another. But because every behavior can be decomposed into a sequence of VAKOG elements and chunks that can be assembled into larger behavioral units, any behavior that is chunked finely enough can be recreated by any person. Moreover, in light of the TOTE model the assembly of target behaviors has a level of flexibility that allows for modifications and work-arounds in the structure of the original behavioral designs (Bandler and Grinder 1975, 1979; Bostic St. Clair & Grinder, 2002; Dilts, 1983; Dilts and DeLozier, 2000 Dilts, Grinder, Bandler, & DeLozier, 1980; Wake, 2010).

There is no such thing as failure, only feedback.

NLP is positive in its outlook. It holds that all communication is a learning process and that when things don't work the way we expect, we then have the opportunity to learn something new. Whenever we seem to fail at a task, we gain the opportunity to find out where we may have erred and can then restructure our approach so that we can do better next time. By taking this perspective, we encounter the world as a continuing adventure with unending opportunity for learning and growth. As a result, we can approach the world with curiosity and the expectation that every problem presents us with new opportunities for growth.

If what you're doing doesn't work, do something else: Insanity is doing the same thing and expecting different results.

In evolutionary biology there is a principle called the law of requisite variety. It states that the organism with the most survival options in a given ecological niche will be more likely to survive than an organism with fewer options. In communications and change work we understand that the person with more options is the one who controls the conversation.

Grinder and Bandler were fond of saying that if you only have one choice you are stuck. If you have two choices, you have a dilemma, but three choices begin to provide real options. Flexibility is a crucial part of expert communication. It is also the correlate of the presupposition that we are responsible for the fruit of our interactions. If we come to the task of communication with no tools, no options, we have no choice but to accept the level of communication that comes to us as a matter of chance alone. If we develop flexible skills we can systematically change our behavior so that we get the results we want.

The twelve-step movement makes use of the same presupposition but states it this way: Insanity is doing the same thing and expecting different results. Watzlawick has pointed out that we often get stuck in a pattern that he calls "more of the same". If something we are doing doesn't work we try to do it louder, more intensely or more insistently. So, we have the caricature of the ugly American who is visiting a foreign country. When he finds that the inhabitants do not speak English—he, of course does not speak their language—he asks his question slowly and more carefully. When the natives prove unresponsive, he repeats the same phrase, a little more slowly, with better diction and a little louder. With each failure, the speaker repeats his query louder and more clearly until he is almost shouting. Finally, he gives up wondering: "What is wrong with these people? Don't they speak English?" This was never the answer. If what you are doing does not work do something different, do anything but what you've been doing (Watzlawick. 1978; Watzlawick, Weakland & Fisch, 1974).

Everyone has or can create the resources that they need to attain their outcomes.

NLP assumes that there is nothing that happens to people on a mental or spiritual level that they cannot learn to handle. Most of the problems that afflict us are rooted in the normal patterns of being alive, being aware and being human. What we have learned we can unlearn or out frame. Needless to say, this does not mean that we all have immediate access to unbounded riches or all of the connections we would like. We were not all born rich or beautiful. Nevertheless, humankind has an extraordinary capacity for creativity and flexibility. We are the only creature that can reprogram our own way of approaching the world, and the only one that can take conscious control of our own personal growth and evolution. In general, NLP provides us with tools for understanding how to do anything that anyone else has done. If you can imagine it, it can be accomplished. If someone else has learned to do it, you can too. Insofar as people are generally not broken, there are no limits beyond the constraints of the physical laws.

On a practical level we can understand this as meaning that every experience that we have ever had can be used as a resource. Modern neurophysiology confirms this as it shows conclusively that memories are recreations of the physiology of the original experience, so that each memory has the potential to make the full biology of the initial experience available. This means that resourceful experiences of love, competence, peace or spiritual awakening can be revivified and enhanced to create new possibilities of experience and action (Erickson, 1954; Damasio, 1999).

People are, for the most part, not broken.

One of the enduring assumptions here in the West seems to be that if someone disagrees with me or encounters the world differently, they are either bad or broken, often a little bit of each. NLP assumes that people who are able to go about life like the rest of us may have problems but those problems do not represent brokenness. More often than not such problems represent poor choices, erroneous habits or other stuck states. The problems themselves are evidence that the organism is fully functional. In this light, addiction is a normal response to the repeated and purposeful use of certain substances to feel good, solve problems or supply other needs. Phobias are simply over-learned responses to fearful or potentially dangerous circumstances. Criminal behavior may be the result of poor choices or an impoverished view of personal options, it is not a defect. As long as a person is physically whole, assuming that they have completed a normal schedule of physical and mental development, we may assume that normal process of learning and change will work for them.

If we break a task down into small enough chunks, any task can be accomplished.

We have already noted the central presupposition that all behavior is composed of ordered sequences of Visual, Auditory, Kinesthetic and Olfactory /Gustatory elements. Implicit in this central observation is the idea that these patterns are common to all fully functional humans and differences in behavior and response can be accounted for by the arrangement of the elements. Meaningful and effective behavior is limited by the ordering of perceptions and actions that result in an effective or recognizable behavior.

Walking is not a random assemblage of movements. It involves specific movements of specific body parts in a definable sequence. The movements related to walking are in turn dependent upon sequences of perceptions that relate to relevant environmental changes and the performance of the task. Although there are variations in the behavior--walking, striding, speed-walking, marching--all involve the same basic syntax of movements. The basic syntactic relations that give rise to a behavior more generally give rise to the more refined sequences that characterize expert or well-practiced behavior.

Modeling consists of finding the patterns and sequences in behavior and making them explicit. NLP holds that every behavior can be analyzed, modeled and taught in terms of specific behavioral chunks (Bandler and Grinder 1975, 1979; Bostic St. Clair & Grinder, 2002; Dilts, 1983; Dilts and DeLozier, 2000 Dilts, Grinder, Bandler, & DeLozier, 1980; Wake, 2010).

Satisfying objections is necessary for lasting change. Respect "resistance."

One of our central presuppositions is that the meaning of your communication is the response that you get. This is usually followed by the assumption that if what you are doing doesn't work, do something—anything—different. Resistance is information; it is feedback. It tells us that in some manner we have failed to understand the client's map. We may not understand the map, we may not appreciate how he values the 'problem' or why it is valuable to him. In general, it tells us that we need to listen more and talk less.

We need to remember that the NLP perspective is radically client focused and that it places an extraordinary focus on the practitioner's capacity for flexibility and curiosity. There is no room for judgmental or authoritarian perspectives—unless that is what the client's map requires. If we are insensitive to client's response, we are not doing NLP.

Chapter Two

The Power of the Name:

Diagnosis—addiction spectrum disorders.

One of the basic presuppositions of NLP is this: the map is not the territory. In the context of professional practice, maps have a way of becoming the territory and it seems to be especially so in addictions studies.

In the West, there is a long history of mistaking names for things. The Pharaohs had secret names known to none but themselves and the gods. To destroy one's name was to destroy the soul and banish the nameless one into empty chaos. Historically, the name of the thing imparted power over it. Our word, *grammar* has its roots in the medieval word *grammarie* which meant magic. One of the great revolutions in science was the Linnaean classification scheme which provided a system for naming living creatures. Classification was often confused with knowledge.

In medical literature, Nosology, the science of naming diseases, is an important discipline. In some sense it is the gate keeper to real knowledge. It is the map upon which medicine often depends for direction. Unfortunately, that map is often a poor one. This is especially so in the field of addictions studies. The following tale may be instructive.

Some years ago, I was a panel participant in a conference on addictions. Two of the other panelists were scientists from a large southern university who announced that they had discovered the genetic markers and the chemical deficits that differentiated between type one and type two alcoholics. The distinction is medically important because the two varieties of the disease (yes, I said disease) have different developmental histories and differing prognoses.

The two young men confidently predicted that since they knew that there were only two kinds of alcoholism and they had essentially found the cure for one of them, they could confidently predict that up to 24 percent of all alcoholism could now be medicated out of existence.

A year later the men returned to report that where they had expected one in four patients to respond; only one in twenty-five did. There seemed to be more going on than just two kinds of alcoholism.

This led me to think about the problem of nosological poverty; our maps don't have enough detail.

Classification systems have levels of detail, the more accurate they are, the more relevant detail they provide. Consider depression.

At the most basic level someone might tell you that they are depressed. For all you know this could be anything from not feeling well, to being sad, to being paralyzed with clinical depression. This is a gross level of nosology. It tells us nothing useful about the problem.

On the next deeper level, a physician might begin to make diagnostic distinctions about the specific variety of depression or mood disorder.

Here, they might discover the following varieties of depression:

- major depressive disorder
- dysthymic disorder
- bipolar disorder
- cyclothymic disorder
- mood disorder due to a general medical condition
- substance-induced mood disorder.
- seasonal affective disorder
- postpartum depression,
- premenstrual dysphoric disorder

Some physicians, notably the GPs who are so happy to open their prescription pads for you, may stop there and fail to note that each of these categories of mood disorder may be subdivided in turn, by 14 possible modifiers and their nearly 200 combinations.

Assuming, however, that our physician is conscientious and finds the correct subdivision of mood disorders, he then begins to prescribe. The problem is not so simple, for no two patients respond exactly alike to the same medication and our physician may find himself rummaging through several hundred kinds of pharmaceutical treatments before finding one that works. At this point, we are working at the level of the pharmaco-phenotype, the level of genetic expression that mediates our responses to drugs. We still have not reached a level of genetic specificity that would allow us to say, as our friends with the cure for alcoholism believed, that we really know the nature of the problem (see Robbins & Gillam, 2011).

In the field of addictions studies, we have at most eight to ten categories and subcategories that describe the problem, but most professionals work with about four. The map is not the territory; that's why NLP works with people and behaviors, not diagnostic categories.

A similar structured analysis of addictions begins with the observation that, for most people in our culture drug problems imply an addiction and alcoholism implies moral failure. When they are equated morally, both are seen as diseases but very different diseases. Alcohol is usually seen as less dangerous.

As an aside, not too long ago a very close friend came and requested help with a severe drinking problem. He was drinking to unconsciousness almost nightly and had extensive blackouts. I arranged for his placement in a medically supervised detox with a short rehab stay afterwards, primarily to give him time to break the pattern and make some plans. He had some family close by who would note his absence and be concerned. With his permission, I called his grandmother and advised her that he would be away for a while at a detox. Shocked, but relieved that he was safe; she asked what he was being treated for. I answered that the problem was alcohol. "Thank God!" she replied, "At least it wasn't drugs."

At the next level of complexity there is some understanding of the basic distinctions between use and addiction, but the distinction is tenuous. More sophisticated analysts will correctly distinguish between use, abuse and dependence or addiction. Highly trained psychologists, therapists, nurses and physicians will make the further distinctions based on the root categories of DSM IV, while others in those professions will make the relatively more fine grained (but largely meaningless) distinctions made in the small print. There, DSM IV differentiates between abuse and dependence, and differentiates them for various drugs and for alcohol. In addition, each kind of dependence or abuse may be current or in remission and if in remission, the pattern (for either dependence or remission) may be full, early partial, sustained, and sustained partial.

Although there is a growing body of evidence pointing to certain genetic dispositions (that are related much more to impulse control than to drug or alcoholism specifically) there is nothing yet in the treatment of addiction spectrum disorders to match the relatively sophisticated pharmaco-phenotyping that is happening with depression.

In general, when we hear the word *addiction,* we are not receiving precise information. Practically speaking, there are four basic levels of problematic behaviors or substance use disorders, which, for the purposes of this study, we will refer to as *addiction spectrum disorders*. This removes some of the unwieldiness of other titles and does not limit the category to drugs and mind-altering substances. Those categories are:

- Addiction
- Dependence
- Abuse
- Casual / recreational use

Even so, there are problems with these categories. Addiction, though current in popular use and in professional literature, is no longer one of the standard diagnostic categories recognized by the American Psychiatric Association. It is assumed to be subsumed under the broader category, *dependence*. As a result, the two are often confused.

Classically, the definition of addiction follows the AMA definition of alcoholism:

> Alcoholism is a primary, chronic disease with genetic, psychosocial, and environmental factors influencing its development and manifestations. The disease is often progressive and fatal. It is characterized by continuous or periodic: impaired control over drinking, preoccupation with the drug alcohol, use of alcohol despite adverse consequences, and distortions in thinking, most notably denial. (Morse & Flavin, 1992)

Højsted & Sjøgre (2007) cite the AMA definition of addiction from Rinaldi, Steindler and Wilford's *Clarification and standardization of substance abuse terminology* (1988):

> ...addiction to opioids "is the compulsive use of opioids to the detriment of the user's physical and/or psychological health and/or social function. Signs of compulsive use include preoccupation with obtaining and taking opioids, apparently impaired control over their use, and reports of craving. Addiction can only be determined by observing these behaviours over time, not on a single event" (p. 492).

According to J. Alan Leshner, past head of NIDA (Leshner, 2005), the most salient features of addiction are loss of control and obsessive preoccupation with seeking, getting and using the drug. These are also possible, but not necessary elements of dependence.

Addictions are typically diagnosed when all of the following nine symptoms have been present for one month or more or have been repeatedly present over a longer time period:

- Taking the drug more often or in larger amounts than intended.
- Unsuccessful attempts to quit, persistent desire, craving.
- Excessive time spent in drug seeking.
- Feeling intoxicated at inappropriate times or feeling withdrawal symptoms from a drug at such times.
- Giving up other things for it.
- Continued use, despite knowledge of harm to oneself and others.
- Marked tolerance in which the amount needed to satisfy increases at first before leveling off.
- Characteristic withdrawal symptoms for particular drugs. Taking the drug to relieve or avoid withdrawal.
- To a large extent, addiction is differentiated from dependence by the persistence of the urge and the loss of control. (Schaeffer, 2005)

Nevertheless, there is a consistent category bleed among diagnostic criteria as used by treatment providers, so that addiction and dependence are often confused. Without rigorous application of the criteria separating them, dependence may look and feel like full-blown addiction. Moreover, dependence and abuse are often confused and, for many people, all use is abuse and all abuse is addictive.

Addiction is experienced by ten to fifteen percent of people who become seriously involved with 'addictive substances'. Many of the rest may suffer from lesser degrees of addiction spectrum problems

A diagnosis of dependence only requires the existence of three or more of the following DSMIV criteria in a 12-month period.

- Tolerance (marked increase in amount; marked decrease in effect)
- Characteristic withdrawal symptoms; substance taken to relieve withdrawal
- Substance taken in larger amount and for longer period than intended
- Persistent desire or repeated unsuccessful attempt to quit
- Much time/activity to obtain, use, recover
- Important social, occupational, or recreational activities given up or reduced
- Use continues despite knowledge of adverse consequences (e.g., failure to fulfill role obligation, use when physically hazardous)   (APA, 1994).

Højsted & Sjøgre (2007) indicate that:

In the International Classification of Diseases (ICD- 10) (World Health Organization, 2003), dependence syndrome is described as "a cluster of behavioural, cognitive, and physiological phenomena that develop after repeated substance use and that typically include a strong desire to take the drug, difficulties in controlling its use, persisting in its use despite harmful consequences, a higher priority given to the drug use than to other activities and obligations, increased tolerance, and sometimes a physical withdrawal state''. The dependence syndrome may be present for a specific substance (e.g. tobacco, alcohol, or diazepam), for a class of substances (e.g. opioid drugs), or for a wider range of pharmacologically different psychoactive substances (e.g. cocaine) (p.492).

In some sense, dependence is sufficient as a diagnostic category to cover both addiction and dependence; however, many professionals insist that it is not. Savage, Joranson, Covington, Schnoll, Heit, and Gilson (2003) indicate that there are three fundamental concepts that lie at the heart of addiction: (1) while some drugs produce pleasurable reward or hedonic impact, critical determinants of addictions are intrinsic to the user, (2) addiction has neurobiological and psychological dimensions—it is multi-dimensional; and (3) addiction is not identical to physical dependence or tolerance (Højsted & Sjøgre, 2007). Addiction is often differentiated from dependence in that the patient often begins with or quickly finds their drug of choice with little experimentation with other drugs and they quickly develop frequent and increasing patterns of use (McKim, 2003; Robinson, 2004).

Højsted & Sjøgre (2007) also indicate that chronic pain patients undergoing otherwise unproblematic treatment with opioids often fulfill at least three of the criteria for opioid dependence. These criteria typically include the development of tolerance, withdrawal symptoms when the medication course is completed, using more of the drugs, more often and over a longer period than they had originally intended and they may have unsuccessfully sought to stop the opioid medications because of increased pain when the dosage was cut.

Beyond a problem with diagnostic criteria and their application, addiction appears to be variable and separable from the various drugs. Some persons are more or less prone to become dependent on some drugs but not others. Others seem to be able to use drugs casually without problems.

According to large government surveys of alcohol users, only about 15 percent are regular dependent drinkers. Among cocaine users, about 8 percent become dependent. For cigarettes, the percentage is reversed. About 90 percent of smokers are persistent daily users, and 55 percent become dependent by official American Psychiatric Association criteria, according to a study by Dr. Naomi Breslau of the Henry Ford Health Sciences Center in Detroit. Only 10 percent are occasional users. (Hilts, 1994, p. C3)

The diagnostic criteria for Substance Abuse Disorder are a bit more straightforward. They include the expression of any one of the following traits in any 12-month period and the patient must never have been diagnosed as dependent:

- Recurrent use resulting in failure to fulfill major role obligations at work, home or school
- Recurrent use in physically hazardous situations
- Recurrent substance related legal problems
- Continued use despite persistent or recurrent social or interpersonal problems caused or exacerbated by the substance
- The diagnosis may not be made if the client has ever been diagnosed as substance dependent. (APA, 1994).

For many years, the standard doctrine of addiction was that drug use inevitably leads to abuse and abuse to addiction. From the classical point of view, addiction was a property of substances, not persons. More recent research suggests otherwise (Bechara, 2005; Leshner, 2005 and see the following chapters).

Phenotypic variation

Robinson, Gillan et al. (2011) argue that that DSM diagnostics are impaired by their superficial nosology and failure to account for phenotypic variations that make each of the diagnostic categories divisible in terms of biological factors. As a case in point they look at compulsivity and impulsivity as biologically based traits in addiction.

Antoine Bechara (2005) suggests that addiction is characterized by defects in competing neural systems, one that inhibits impulsive behavior and another that drives the urgency of cravings. The first is dominated by frontal inhibitory structures and the latter by the evaluative functions of subcortical structures including the amygdala and the ventral striatum. This matches the distinctions between compulsivity and impulsivity that was later made by Robinson and Gillan (2011) and also comports with Cloninger's (1987) differentiation between the impulsive drug seekers and the relatively compulsive alcoholics characterized by loss of control (see below).

Bechara, with others, points out that one of the important things that an adequate approach to addiction must consider is why it is that so few people who experiment with drugs end up with serious addictions. He suggests that there may be predisposing factors that lead some groups to be more susceptible to addiction than others. As a result, he suggests that addictions must be treated in a highly individualized manner (Bechara, 2005; McKim, 2003; Robinson & Berridge, 2003).

Inhibitory responses related to the exercise of self-control include the Ventro-Medial Prefrontal Cortex (vMPFC) and support structures including the dorsolateral prefrontal cortex, the insula and the parietal cortex. Addicts often respond to tests of impulsivity much like patients who have damage in these areas. He reports Damasio's study of Nicolas Gage and the radical change in his level of impulsivity occasioned by damage to these structures. Strikingly the insula, where bodily sensations are associated into an ongoing sense of self has been crucially related to addiction, Naqvi and coauthors have shown that addicts have decreased cortical tissues in the insula. They also report rat studies in which ablation of the insula ended various addictions. (Naqvi, Rudrauf, Damasio, & Bechara, 2007; Naqvi & Bechara, 2009). Raine and colleagues (Raine, 1993; Raine, Buchsbaum, & la Casa, 1997) report that an overwhelming plurality of violent criminals have suffered head wounds also suggests that pre-existing frontal damage may contribute to addiction vulnerability. Multiple authors have also pointed to defects in frontal function as a crucial element in the creation and maintenance of addiction spectrum disorders (Chambers et al. 2007; Feil et al, 2010; Redish et al., 2008).

According to Bechara, intact functioning of the vMPFC, anterior cingulate, insular cortex, dorsolateral prefrontal cortex and lateral orbitofrontal/inferior frontal gyrus are crucial for making responsible decisions that involve weighing future consequence and rewards. These, however, are the very centers that are compromised by addictive processes. Unfortunately, because the impact of substance abuse so closely parallels damage to these areas, it is almost impossible to determine whether damage to functioning in these areas is a precondition for full-blown, downward spiraling addiction, or the result of the drug use itself (Bechara, 2005; Chambers et al. 2007; Feil et al, 2010; Redish et al., 2008).

Bechara reports a series of studies using the Iowa Gambling Task to compare the responses of addicts to frontal lobe patients with damage to the vMPFC. The Iowa Gambling task simulates card games. In one condition players obtain large early rewards in the form of money credits but eventually lose out. In another condition, players gain small credits consistently and despite their small size, end up with considerable winnings. Normal players quickly learn that the condition that provides the early large rewards is a losing proposition and switch to the other consistently-paying game. Frontal lobe patients and up to two-thirds of addicts—even those in remission—are unable to switch tasks, even though they can articulate the difference between the two games. The results have been duplicated many times (Bechara, 2005; Bechara & Damasio. 2002; Bechara, Damasio, & Damasio, 2000; Bechara, Damasio, Damasio, & Lee, 1999; Bechara, Dolan, & Hindes, 2002; Feil, Sheppard, Fitzgerald, Yücelc, Lubman, & Bradshaw, 2010).

In his review, Bechara (2005) reports that 63% of addicts responded like those with frontal defects, and 27% of the normal controls also responded like frontal lobe patients. He suggests that the 27% of normal controls who responded like addicts and persons with damaged vMPFCs may represent a group at risk for full-blown addiction. He continues by noting that the addicts who responded as if they had frontal lobe damage could be divided into two groups based upon electrodermal responses during the experiment. Half of them showed an abnormally high level of excitement in the presence of reinforcement. That is, their deviation from the norm, their lack of impulse control, was more related to their excitement over the reward than to specific problems with the inhibitory structures in the frontal lobes. These results immediately suggest that under the rubric of addiction, we are potentially dealing with three distinct subpopulations: those who's addiction mimics or is driven by frontal damage, those whose addiction is driven by an oversensitivity to reward and a third population with no apparent predisposing conditions. To this list we may also add addicts with abnormally functioning insular cortices.

In alcohol addiction, Cloninger (1987) identified two distinct types. Type one alcoholism typically starts after age 25, is characterized by binging and violence and is often environmentally driven. The central issue for type one drinkers is loss of control. Type two begins before age 25 and is characterized by the inability to remain abstinent by age 25. He also proposed that his genetic studies differentiated between the roots of alcohol seeking and loss of control.

Cloninger, hypothesized that alcohol addiction developed and was maintained by genetic effects in three systems: harm avoidance, novelty seeking, and reward dependence. In a work that anticipates later descriptions of the mechanisms of addictions, he proposed that: the harm avoidance system was heavily dependent upon prior learned associations mediated by serotonergic connections and the novelty seeking system was linked to the midbrain dopamine system. Reflecting on the often reciprocal balance between serotonin and dopamine, he suggested that alcohol seeking behavior, characteristic of type two drinkers, was dopamine driven while loss of control, the major feature of type one drinkers, was driven by defects in the serotonergic system. He related reward dependence to defects in the noradrenergic system that resulted in increased habituation to the acute effects of alcohol. Other researchers (Robinson & Gillam, 2008) have related the reward dependence trait to the transition from impulsive drug use to habitual drug use.

Type one individuals, whose alcoholism is rooted in loss of control, tend to be anxious and dependent, they are people pleasers who are cautious and conservative. Type two individuals, who actively seek alcohol and are unable to abstain, are often anti-social types who seek novelty, unconcerned about consequences and self-directed.

Reddish et al. (2008) classify ten discrete behavioral categories where decision making may be compromised in addictions. This suggests that any useful diagnostic might need to consider where in the decision to use or to continue using any of the ten defects or their multiple interactions provides meaningful information.

For simplicity's sake, focusing only on the work of Bechara and Cloninger, we find predisposing factors that may provide insight into a more precise nosology of addiction. Examining drugs that are often held to be polar opposites in their superficial effects, both authors find that addictions may be characterized by defects in inhibitory control, reward sensitivity/novelty seeking, and abnormal levels of behavioral perseveration. Thus, beyond the eight to ten dimensions of the addiction spectrum specified by the diagnostic manuals, we can add the permutations of at least three dispositional factors that vary independently (and may themselves consist of separable components (Robinson & Gillam, 2008) and leave us with a minimum of one-hundred different species of drug problem without reference to the substance abused.

Further complications

We have already noted that the classical treatment model holds that addiction is a property of the drug and that because *The Drug* is the problem, any use leads to abuse and abuse inevitably leads to addiction. While technically in error, this is the presupposition of many treatment providers. As a result, the focus of treatment often becomes the drug and not the person.

Further, the idea that addiction is a primary, lifelong, and often fatal disease continues in the DSM IV criteria where dependence is never cured but may be in remission—once an addict, always an addict. This is despite multiple studies that indicate that a large proportion of people, who have abused legal and illegal substances at apparently addictive levels, can return to casual use without problems (Miller, 2004; Peele, 1987; Peele, Brodsky & Arnold, 1992).

Problems arise once more when we understand that all levels of substance use up to and including dependence may be contextually determined. These cases would include people who cannot refrain from smoking while drinking, who cannot refrain from using drugs or alcohol while in the presence of certain people or places. There are other people for whom a long period of problematic use at addictive levels may be resolved by a physical move, encountering and joining an appropriate group, getting religion or finding a spouse or *meaningful* occupation.

Two other dimensions of addictive spectrum disorders are the motivation towards change and the problem of judicially mandated treatment. It is important to realize that someone who is personally, deeply and intrinsically motivated to change is a very different person from the person who is not. In classical addictions literature the point was strongly made that intrinsic motivation was the sine qua non of treatment success. In more statistically driven approaches, however, it was seen that treatment completion was highly correlated with treatment success and that nothing provided compliance better than a court mandate.

In general, if we can provide the client with a motivation to change, the task of change becomes much easier. The Stages of Change Model describes the process and motivational interviewing provides a significant tool for creating such motivation to change. Nevertheless, once someone has decided, *in a fundamental manner*, that change is really important, almost any intervention can work.

For most of us who work or have worked with addiction spectrum problems, one complicating problem is judicial diagnosis. Like iatrogenic problems in medicine, drug problems may be created by a judge's mandate. First, the impact of the judicial determination may be damaging in itself. Second, and perhaps more importantly, finding themselves in hopeless circumstances, some persons known to the author have begun substance abuse or changed to more serious drugs because of the circumstances to which the court order exposed them. The literature on labeling theory suggests that this may be a more likely problem than we would like to think (Becker, 1963).

In almost 30 years of experience in the criminal justice system the author has seen hundreds of persons sent for addiction treatment, often including detoxification, because at some time in the past they used drugs recreationally, had a problem with drugs, or talked to someone who was using drugs. Even though, on an objective level, these people have no problem with alcohol or drugs, the judicial fiat makes them our responsibility and appropriate treatment plans must be developed.

Therefore, when considering treatment we must determine whether the person is addicted or dependent, an abuser or a casual user. We must discern whether the problem appears to be chronic and recurring, situationally bound or socially motivated and whether the client is intrinsically motivated to change or not. We must further determine whether the mechanism that maintains the problem is genetically rooted, developmentally based, or arising out of other pre-existing conditions.

Briefly, we can say that, on the grossest level, most of the people who come for treatment for addiction spectrum disorders will fall into one of the following categories:

- Judicially mandated because of association, suspicion or past use—no present problem
- A casual user or abuser referred by friends, family or the Courts who uses alcohol or illicit substances but has no problem with their own use patterns. This is substance abuse disorder diagnosed by interference with personal choice.
- Persons who have lost control over substances or behaviors in specific contexts—the problem does not exist except at certain times, with certain people or in specific places. These people may meet diagnostic criteria for abuse or dependence but the problem is always limited by context.
- Persons using opioids under medical supervision who experience no substance related problems but nevertheless technically meet diagnostic criteria for dependence.
- A person who has experienced increasing loss of control over substances or behaviors and who meets diagnostic criteria for substance abuse disorder and the problem is not contextually bound. There are motivated and unmotivated types.

- A person who is dependent upon a substance or behavior, but does not experience obsession or craving. There may be withdrawal or not but once the problem is over, it is over. They may in fact be able to use the substance without problems in the future. There are motivated and unmotivated types.
- Persons who experience classical and chronic addictive symptoms with recurring bouts of heavy use, multiple failed attempts to quit and patterns of progressive use with decreased effect. There are motivated and unmotivated types.

In the end, we must agree with Bechara that each addiction must be dealt with as a unique manifestation of the interaction between an individual and a substance or behavior.

Chapter Three

Three Important Studies

The classical addiction literature and the preachments of law enforcement tell us that addiction is a property of mind-altering substances. Alcoholics are fond of quoting Bill W. or Doctor Bob as saying that drugs and alcohol are cunning and wily foes. Indeed, the entire scheme for the legal classification of drugs is determined by whether they have a potential for abuse and whether or not they have –legislatively acknowledged— medical use. Please note that the medical application is not a scientific determination, but a legislative one.

Schedule One drugs are defined at 21USC 812(b) (1) and list drugs or other substances which the lawmakers have determined to have a high potential for abuse. According to those same lawmakers, the drug or substance has no currently accepted medical use in treatment in the United States. They have also decided, often independently of evidence to the contrary, that there is a lack of accepted safety for use of the drug or other substance under medical supervision.

Marijuana is included in Schedule I, the most dangerous category of drugs. Attempts to find a lethal dose (LD/50) for this substance have failed time and time again, while at the same time, its medical uses keep expanding. In a finding of fact regarding a petition to reclassify marijuana as a Schedule II drug—legal for restricted medical use—Administrative Law Judge Francis Young (1988) found that "A smoker would theoretically have to consume nearly 1,500 pounds of marijuana within about fifteen minutes to induce a lethal response. ... In practical terms, marijuana cannot induce a lethal response as a result of drug-related toxicity" (Sec. VIII, para.8-9). Surprisingly, alcohol and tobacco are absent from the list while marijuana is present. It should also be noted that drugs that are presently prescribed in other countries are included in the forbidden category.

As a reflection of the logic of the classification scheme, it should be noted that cocaine is a Schedule II drug. If we consider the invalid argument that marijuana is a gateway drug that often leads to the use of cocaine, we are caught in the dilemma that marijuana, the legislatively more dangerous drug, might lead to the use of cocaine which, by legislation, is less dangerous. These are neither scientific nor rational determinations.

As noted above, addiction does not appear to be a single phenomenon. Its rates change from drug to drug. According to J. Alan Leshner, it appears to be a brain disease. The following studies may be enlightening.

STUDY NUMBER ONE, RAT PARKS.

Alexander, Bruce K. Beyerstein, Barry L., Hadaway, Patricia F., & Coambs, Robert B. (1981). Effect of Early and Later Colony Housing on Oral Ingestion of Morphine in Rats. *Pharmacology, Biochemistry & Behavior,* Vol. 15. pp. 571-576.

In this study, rats were raised either in cramped single cages or spacious rat parks with other rats, room for exercise, potential mates and places to explore. The aim of the study was to determine the effect of environment on opiate addiction. The authors indicate that, in certain circumstances, animals can be trained to drink morphine water in preference to plain water and to self-inject morphine through implanted catheters. The evidence from these studies has been taken to mean that animals have a constitutional affinity towards opiates and their effects. That is, rats, could become addicted to opiates. In this study, and in other studies that preceded it, researchers found that addicted rats raised in or moved from cages to 'rat parks' drank far less morphine water than did their caged brethren. The change in behavior held both for rats in which an addiction to morphine had been established as well as for rats that had no experience with morphine water (Alexander, Beyerstein, Hadaway, & Coambs, 1981).

The researchers suggested that the avoidance of morphine by the rats raised in or moved to rat parks (colony rats) could be explained by the fact that the effects of morphine interfered with species specific behaviors which are strongly evoked when rats live in colonies. These behaviors included nest building, mating and fighting.

These expectations were suggested by previous research that had shown that small amounts of morphine interfered with self-reinforcing or autotelic behaviors including sexual activity and other socially oriented responses. A secondary hypothesis suggested that the effects of morphine had a calming effect on isolated rats and so reinforced their morphine seeking and using.

In the experiment, 16 Wistar rats (a standardized breed often used for such experiments) were separated into two groups. For their first 60 days after weaning, one group was raised in standard wire laboratory cages (7x7x9 inches) while the other group was raised in 30-foot square, open-topped pens equipped with cedar shavings, empty canisters and small boxes in which the rats could hide and nest.

At 65 days, one-half of the caged rats were transferred to the pens and one-half of the rats from the pens were moved to the cages. This created four groups of rats: rats that had spent all of their lives in either in the rat parks or the cages and those who had begun their lives in either condition and moved to the other.

After a 15-day accommodation phase, all of the rats were given 24-hour access to both plain tap water and a morphine sugar solution with strengths of the morphine solution varying across seven stages.

In stage one; the rats received water and a water sugar combination to test whether housing conditions created a preference for sweet things. In the next stage, the rats were allowed access to tap water or a solution of quinine and sugar. This tested for a preference for bittersweet flavors and provided a sensory experience that humans cannot distinguish from morphine and water. A previous test of rat preferences for these solutions found that rats drank them in roughly equal amounts.

During the next four stages of the experiment, the rats were allowed access to either water or decreasingly powerful solutions of morphine and sugar. A final stage of the experiment repeated the water sugar/quinine solutions of the initial phase.

It was found that none of the rats drank much of the strongest morphine solutions in either condition (cages or rat parks). It was believed that the solution was just too bitter. Male rats who were living in the rat park at the time of the study took far less of the other morphine solutions than did the rats in cages but both drank equal amounts of the control solution. At one level of morphine concentration, caged rats took 16 times as much morphine as did colony rats.

Early environment alone did not predict morphine use, but rats who had been caged in early life and were then moved to the rat parks were more likely to choose the morphine solutions than those who had been raised in the rat parks. In general, rats that were tested while living in the rat parks, no matter which condition they began in, were less likely to use morphine than the rats that were tested while living in cages. Although there were some differences in female responses, none were significant and gender was not found to be an important influence in the study.

The authors determined that the main effect was contributed by some difference between the cages and the rat parks. As hypothesized, the experiment seemed to reinforce the idea that open spaces, the opportunity for sexual behavior and other social interactions were more rewarding than the effects of opiates.

This study suggests that there may be natural preference hierarchies that are organized in terms of the opportunities that they afford. The rats that were given access to an environment that tended to support self-reinforcing 'instinctive' behaviors were less likely to choose morphine solutions than were the rats that had no such opportunities. The study also suggested that early stress made the choice of opiates more likely for rats that had moved to the colony situation; all of the rats living in the positive environment were less likely to choose opiates than the rats in cages.

This famous study is often cited as a central pillar in the argument that addictions are not about the drugs... whether you believe that animal studies apply directly to humans or not, it is at least highly suggestive. Please consider that on a physiological level, Nature is highly conservative. We share upwards of 96% of our genetic makeup with chimpanzees (nevertheless, that represents something on the order of 40 Million individual coding differences) and 40% with the rat (Henderson, 2003; Rat Genome Sequencing Project Consortium, 2004). This suggests that there must be some overlap.

The most important piece of information provided by this study may be that addictions are controlled less by the drugs than they are by the opportunities that an individual perceives beyond the drugs. If there are options of value to the individual, s/he may be less likely to begin or to continue drug consumption.

Although the idea of human (and to some extent, animal) instincts has passed from favor, we can still understand humans as having needs and tendencies to respond. Like most organisms we respond to pleasurable stimuli—tastes, petting, variety, sex, warmth, etc. We respond to the same kinds of conditioning behaviors as do other mammals and, it would appear that these preferences or needs are arranged hierarchically under the influence of immediate states of deprivation and satiety.

STUDY NUMBER TWO, VIETNAM VETERANS

Robins, Lee N., Davis, Darlene H. & Nurco, David N. (1974). How Permanent Was Vietnam Drug Addiction? *American Journal of Public Health. Supplement*, Vol. 64, December, 1974

During the Vietnam War, drug use ran rampant among U.S. service members. By the spring of 1971, it was estimated that almost half of all GIs had been using heroin at addictive levels during their tours of duty. In response, the military began urine screens of returning vets and determined that the GIs who tested positive for illegal substances would be detained in Vietnam for six or seven days until their urine samples no longer tested positive. The testing found that despite the warning and the possibility of their return being delayed, five percent of those awaiting return to the States tested positive for recent drug use.

Even before the testing program, there was a great deal of concern that if a large number of GIs were indeed addicted to heroin or other drugs, they might have difficulties finding work and might turn to crime in order to purchase the much more costly American heroin. As a result, the White House Special Action Office for Drug Abuse Prevention (SAODAP) authorized a study to determine how many servicemen would require treatment for addiction, how these men were to be identified and what specific treatment and social services they would require. The study was also aimed at determining the natural course of substance abuse and addiction, especially in circumstances where drugs were readily available.

In September of 1971, approximately 13,700 GIs returned from Vietnam. A simple random sample of 470 of this group was identified for the study. Among those same returnees, a group of 1,400 had tested positive for illicit drugs (10% of the larger sample). From these, a simple random sample of 495 individuals was selected. Interviews and urine samples were taken for the study participants during the period between eight and twelve months after their return. Their military records were also examined and any claims that they may have made for veterans benefits were checked.

The researchers indicated that:

> Military records were obtained for 99 percent; and VA claims records for 22 percent. Interviews were obtained for 95 percent and, of those interviewed, 98 percent of the General Sample and 97 percent of the Drug Positives provided urine specimens. Since the rate of interview was 97 percent of the 466 surviving members of the General Sample and 95 percent of the 493 surviving members of the Drug Positive sample, and since over 90 percent of every subgroup defined by race, age, rank, or type of discharge yielded interviews, unbiased estimates of responses by both drug free and drug using veterans were virtually insured. (p. 39)

Study participants were asked about their observations of substance abuse in Vietnam, their personal opinions about how to handle the problem and their own personal drug-use history. They were also questioned about "drug and alcohol use, family problems, marital history, social relationships, school difficulties, job, arrests, depressive symptoms, psychiatric treatment, and disciplinary actions" (p.39). Drug histories were divided into five time periods: those whose history began before service, those who started in service but before Vietnam, those starting in Vietnam, those who started using while still in the service after Vietnam and those who had begun using since discharge.

All of the answers were checked against military history and other documents. Interviewers did not know whether their subjects were drug users or not and the informants were not told that their answers would be checked against military and other records. When checked, 97 percent of soldiers who had tested positive for drugs in Vietnam admitted their heroin use while in Vietnam. Eighty-four percent of the sample acknowledged that their samples—taken in Vietnam—had been found to be positive for heroin. Of those interviewed, 81 percent had left the military and 75 percent had returned to their hometowns.

Despite the original urine tests indicating that only five percent of GIs in Vietnam had been using drugs, 43 percent of the general sample admitted drug use while in Vietnam. About 46 percent of those who reported using drugs reported addiction and 23 percent of the users tested positive upon their departure from Vietnam.

One striking finding was that even though 43 percent had indicated drug use while in Vietnam, the rate dropped to 10 percent of the general sample after their return to the States. For addiction, the number of those reporting addiction dropped from 23 percent in Vietnam to 7 percent in the states—one percent of the general sample. The self-report numbers were confirmed by the frequency of positive urines in the sample population.

Heroin was the most frequently reported drug abused either while the GIs were in Vietnam or after their return to the States. Of those who acknowledged substance use, 79 percent of informants acknowledged heroin use in Vietnam and 74 percent indicated the same preference after their return.

A significant part of the results found that the rate of drug use and addiction reported after the return from Vietnam was essentially the same as the rates reported before service. Nevertheless, for those who did continue using, their use was more regular, more consistently heroin-based and more often addictive.

Despite the increased intensity of addictive problems for those who continued to use drugs, the study reports that the 95% remission rate that was found among the GIs (the rate decreased from 20% of the sample in Vietnam to 1% after their return), is without parallel in the study of addictions.

There were three important subgroups in the sample: 55 percent of the general sample who were abstinent before, during, and after service; three percent of the sample who were heroin abusers before during and after service; and 27% who had never used before Vietnam, used in Vietnam but never used after returning to the States. In the last group, however, some (17 %) had used other drugs before service and continued to use them afterwards, others (9 %) were using drugs other than heroin that they had first used in Vietnam and three percent began using other drugs after returning from Vietnam. Nevertheless, all had stopped using heroin.

The results indicate that many soldiers who had begun using drugs before Vietnam stopped after leaving Vietnam, another group that began using in Vietnam continued to use after their return home. The authors note:

> This raises the possibility that Vietnam may not only have introduced some soldiers to narcotics for whom drugs will be a long term problem, but also may have hastened the dropout from use for some pre-Vietnam users, perhaps by

speeding up the addiction process or by their witnessing other people's problems there. Of those who had used narcotics before Vietnam and continued using them there, 75 percent quit by the time they left. Of those who used narcotics for the first time in Vietnam, 80 percent quit on or before departure. There is remarkably little difference in rates of quitting between these two groups. (p. 40)

One of the crucial questions raised by the study was whether or not the men studied actually were addicted to heroin. Those who had reported drug use while in Vietnam were asked what drugs they had used, how many times they had used them and for how many months they had used them more than weekly. They were asked about the existence of withdrawal symptoms: whether they had experienced them, how many times, what they were and how long the symptoms lasted. Eighty-percent of those who were questioned indicated that they had used heroin and /or opium regularly for more than six months and that all but two percent of these had reported withdrawal symptoms. In 97 percent of the cases reporting withdrawal, the symptoms lasted for more than two days.

To ensure that the observed remission rates represented changes in the behavior of true addicts, the researchers examined the results for subjects who had the following signs of addiction: they were still using when they left Vietnam, they had used frequently for more than a month, they considered themselves addicted and they had serious long lasting withdrawal symptoms. They report, "Of this group, only 9 percent reported readdiction in the 8 to 12 months since their return to the United States, and 57 percent said they had not used narcotics at all since they came back" (p. 42).

Our culture has certain expectancies about addiction and the drugs associated with them. In apparent contradiction of these assumptions we find the record of thousands of GIs who were addicted to heroin returning to their homes and, for the most part, leaving the addictions behind. This represents a 95% remission rate with no treatment when the best current treatments seem to top out at about 30% (Di Clemente, 2007).

It appears that addictions may be contextually bound independent of the substances involved. Heroin is normally held to be highly addictive; it is famous for its withdrawal syndrome. The modern myth of the addict as monster is usually based on the idea of the ravening heroin addict. Nevertheless, these men returned from Vietnam in their thousands and for the most part walked away from drugs.

Peele, Brodsky and Arnold (1992), after reviewing the same data suggested that upon their return home, the reassertion of normal roles, the values of family life and their reconnection to the community were much more biologically relevant than the stress of a war that had been left far behind. Among those for whom the roles and values were already strong, the transition was easy. For others, it was not so easy. They suggest that helping people to find or re-find their place in the world would have a significant impact on the thing we call addiction.

There is another striking fact that emerges from these numbers. While addiction rates returned to their pre-Vietnam levels, not all of the post- Vietnam addicts were pre-Vietnam addicts. Many of those who began their tours of duty already addicted, returned home free from addictions. An equal number, who had never known substance abuse before Vietnam, returned to America addicted. This suggests that contexts change observed rates of addiction. Some people gain sober contexts while others lose them to addictive contexts.

Gray (2008, personal communication) reports a client who served in the army during the Vietnam War. During his six-month stay in Vietnam, he became heavily addicted to heroin, using it daily during most of the period. No one had told him about withdrawal and, on the flight back to the US, he experienced what he thought was a bad case of the flu. He reported that when he saw the hills of Oakland, CA, his symptoms disappeared and he remained drug free—not even thinking about it—for eight years. After eight years, an old acquaintance offered him some heroin. He took it and was quickly re-addicted.

STUDY NUMBER THREE PAIN, OPIOIDS AND ADDICTION

Colleau, Sophie. (1998). Research in Cancer Pain and Palliative Care. Pain: Opioid Use and the Incidence of Addiction. *Cancer Pain Release*. Volume 11, No.3.

In 1998, the World Health Organization published a survey of research regarding the use of opioids for palliative care in pain treatment. Colleau (1998), the editor of the survey, examined 11 studies regarding the incidence of addiction in medically supervised palliative care contexts. She noted that:

> Overall, these surveys provide evidence that addiction is exceedingly rare during long-term opioid treatment of cancer pain and does not commonly occur among patients with no history of abuse who receive opioids for other medical indications.

Colleau and Joranson (1998) reported results from their examination of studies accounting for 24,000 patients with no previous history of substance abuse who received opioid treatment for pain. The study found that only seven out of the 24,000 became addicted. They also reported that cancer patients who received long-term opiate treatment could stop the drugs when the pain ended. That equates to an addiction rate of three one-hundredths of one percent (0.00029).

Incidence of addiction from Colleau's summary				
Author	Date	Sample size	Care context	Addicted / dependent
Porter & Jick	1980	12,000	Pain control	4
Medina & Diamond	1977	2,369	Headache Center	2
Schug, Zech, Grond, Jung, Meuser, & Stobbe	1992	550	Cancer treatment	0
Perry & Heidrich	1982	10,000	Burn center	0
Portenoy & Foley	1986	38	Severe non-cancer pain	2
Zenz, Strumpf, & Tryba	1992	100	Diverse pain syndromes	0

The above table summarizes the data for those studies providing numerical data.

We are once again confronted with data that seems to contradict everything that we know about addictive drugs. Drugs should exercise complete control over these people. Some of them used opiates like morphine for years but after mild withdrawal, they walked away. The only way that this can be made to make sense is if we again evoke the idea of context. In pain treatment, pain becomes the context and the meaning of drugs becomes only—what takes the pain away. When the pain is gone, the context for drug taking is also eliminated.

This might be illustrated by place preference studies. One of the important ways that researchers assess addiction in animals, other than humans, is by place preference. Repeated research has found that if you give 'addictive drugs' to an animal in a certain place and that animal becomes addicted to the drug, the animal will show a preference for that place. That is, he will return to that place as if hoping to receive more. An extension of place preferences is the observation first made by Seigel that addicts who use drugs repeatedly in the same place build up a tolerance for the drug in that place so that they can actually use more of the drug in that circumstance than they might in another. In both cases, the drug response is modified by a place or the meaning of a place is modified by the drugs. In the case of pain, pain-as-place, the pain changes the meaning of the drug. Is it possible that for the GIs returning home, the change of place changed the meaning of the drug (Mucha, van der Kooy, O'shaughnessy & Bucenieks, 1982; Seigel, 1982, 1984)?

Chapter Four

Mechanisms of Motivation and Reward:

When we think of addictions, we think necessarily of drugs with an overwhelming power to dominate consciousness and destroy lives. While this is in some measure true, it is also an exaggerated account that is based more on propaganda and fear than it is on facts. Despite years of doctrine holding that drug use inevitably leads to addiction and that abuse is a stage in an inevitable downward spiral, there is growing evidence that only ten to fifteen percent of all persons who use drugs become addicts (Bechara, 2005; McKim, 2003; Robinson & Berridge, 2003). Further evidence indicates that many people with significant addictive careers spontaneously turn away from drugs with and without treatment (Robins, 1973; Robins, Davis & Nurco, 1974; Robins, Helzer & Davis, 1975; Waldorf & Biernacki,

1979; Lopez-Quintiero, Hasin et al., 2011). Other evidence suggests that most addicts begin with their drug of choice without the intermediary of so-called gateway drugs and that addicts usually begin with heavy use and proceed rather quickly —over the course of about a year of heavy use—to addictive patterns (Canales, 2005; Inaba & Cohen, 2007; McKim, 2003). This suggests that addiction is much more about the person using the drugs than it is about the nature of the drugs themselves.

In fact, we will argue that the phenomenon of drug addiction has multiple dimensions. These include the properties of the drug itself, the contexts of use—including social, environmental and personal contexts, the pattern of use leading to addiction and characteristics of the individual who becomes an addict.

THE ADDICTIVE PROPERTIES OF DRUGS OF ABUSE

THE ACTION OF DRUGS ON BASIC NEURAL FUNCTION

The truism that addiction is about the drugs reflects the physiological reality that certain kinds of drugs have very specific kinds of impacts on the human brain. Drugs that are characterized as addictive have the capacity to 1) imitate natural neurotransmitters and neuro-modulators and 2) affect the transmission of dopamine in the motivational centers of the brain, sometimes directly and sometimes indirectly.

The nervous system is composed of nerves that communicate with one another across fluid-filled gaps or synapses. The message from one neuron to another is passed across the synapse by neurotransmitters and the efficiency of that transmission is modified by other chemicals at sites away from the synaptic endings called neuro-modulators. Each kind of nerve or neuron typically uses one basic kind of neurotransmitter to communicate with the other cells in its network

Neural messages are currently thought of as electro-chemical reactions that are transmitted from the neuron's cell body, down the length of the axon to the synapse, where they cross the synapse and stimulate the dendrites of the next neuron. If the amount of excitation at the dendritic terminals is sufficient, and of the right quality, the cell body of the next neuron initiates a message which is likewise transmitted down its axon to its own synaptic endings.

One useful metaphor that has been used to explain the action of neurotransmitters is based on the image of keys and locks. Intersynaptic transmission can be thought of as if the originating neuron sent a flood of keys across the synapse to the dendritic surfaces of the next cell. On the terminal surfaces of the dendrites are receptors for the chemicals, which, like locks, only respond if the right key is transmitted. If enough keys open enough locks, the message is transmitted to the next cell body.

After a neuron has fired and has sent packets of neurotransmitters across the synapse; in order for the neuron to continue functioning effectively, the neurotransmitters must be eliminated from the receptors (or locks) to which they have connected and from the intersynaptic fluid, thus clearing the signal. This happens through the action of several mechanisms. In some cases there are chemicals that break down the neurotransmitters into their chemical components, thus ending their capacity to stimulate the next nerve. In other cases, the neurotransmitters are reabsorbed by the presynaptic neurons for later reuse by the cell. This likewise ends their ability to continue stimulating the following neuron. Drugs of abuse affect this system in several ways.

Some drugs of abuse imitate the action of the neurotransmitters themselves. In so doing they provide a much more powerful stimulation of the neurons next in sequence than natural processes ever could. In effect, they flood the synapse with an imitation of the appropriate neuro-chemical and cause it to fire more intensely than it could on a natural level. In a related process, the drug forms a version of the key that, after opening the lock, sticks in the lock and resists reabsorption or destruction. As a result, the neural message is continually activated, often for much longer and much more powerfully than usual. This kind of action is called agonism. The chemical in this case is called an *agonist*.

Other drugs block the action of the neuro-transmitter at the dendritic end of the synapse. In this case, it is as though the key were broken off in the lock before the lock could be opened. It prevents the use of the synapse and prevents neural transmission. This is called antagonism and this kind of drug is called an *antagonist*.

A third mechanism of drug action is reflected in the capacity of the substance to block the breakdown or reuptake of the neurotransmitter. By allowing excess amounts of the neurotransmitter to remain in the intersynaptic fluid, the cells continue to stimulate each other and the original message continues to be sent for a longer period or more intensely than the natural state of the cell would allow. Those that block reabsorption are called *reuptake inhibitors*. Those that block the breakdown of certain classes of neurotransmitters are called monoamine oxidase inhibitors (MAOIs)

Certain other drugs work as neuromodulators and either increase or decrease the ability of the nerve to produce or respond to their native neurotransmitter. Their action typically affects the electro-chemical properties of the cell membrane along the axon.

Several major neurotransmitters are significantly impacted by substances of abuse. These are dopamine, serotonin, acetylcholine, epinephrine, norepinephrine, GABA and glutamate. Neuromodulators similarly impacted include the endorphins and anandamide.

Because all of these chemicals can change the way the nervous system responds, all of them can significantly alter consciousness. However, knowing that a certain neurotransmitter or neuromodulator is involved is not enough. Where these effects happen in the brain, which networks, circuits, pathways, nodes, lobes and tissues are affected is what gives these changes their significance.

Cocaine typically acts as a reuptake inhibitor for the neurotransmitters dopamine, serotonin and norepinephrine. It also blocks the presynaptic transporter protein for dopamine (the protein that moves dopamine from the cell body to the synapse). These actions result in significant increases in the intersynaptic concentrations of the neurotransmitters which create unnaturally powerful and long lasting neural excitation. Methamphetamine acts by stimulating excess expression of dopamine and by reversing dopamine reuptake in the presynaptic neuron. By these mechanisms it increases the levels of neural stimulation. Regarding addiction, the primary loci of their effects are in the mid-brain dopamine system (the reward pathway) and in hippocampal circuits (where memories are created) (Centonze, D., Picconi, B., Baunez, C., Borrelli, et al., 2002, Hyman, Malenka & Nestler, 2006).

MDMA affects the neuro-transmitters serotonin, dopamine and norepinephrine but the bulk of its subjective effects are due to the action of serotonin. The major effects of MDMA are typically observed in the ventral Tegmental Area (VTA)--the source of dopamine for the midbrain dopamine system--and in the nucleus accumbens--the brain's reward evaluator and one of its pleasure centers--(Bankson, & Yamamoto, 2007; Liechti & Vollenweider, 2001).

Heroin works on multiple neurotransmitter systems. Heroin binds to receptor sites for endogenous opioids and also reduces the production of GABA. GABA normally inhibits the production of dopamine in the VTA but when GABA levels are reduced, dopamine levels are increased (Koob, 1992; Hyman, Malenka & Nestler, 2006).

The action of alcohol and benzodiazepines is closely related to the functioning of the GABA system.

While these relationships explain some of the actions and subjective effects of addictive substances, they do not explain the mechanism of addiction itself. The mechanisms of reward, dependence and addiction all appear to be related to the actions of dopamine and opioid receptors in the midbrain dopamine tract.

Addiction, Dependence and the midbrain dopamine system

During the 1950s work by the neurophysiologist James Olds (Olds & Milner, 1956) led to the discovery of certain brain areas that, when stimulated electrically, gave rise to experiences of ecstatic, pleasurable states. Both humans and other organisms, when allowed to self-stimulate with electrodes implanted in these "pleasure centers," were reported to display behaviors similar to addictions; they would self-stimulate at high rates while neglecting other sources of reward. Rats were widely reported to have produced thousands of responses per hour and would neglect eating and drinking in favor of the electrical stimulation of the pleasure centers (Hyman et al. 2006; Schultz, Dayan & Montague, 1997). As time went on, researchers discovered that the electrodes in Olds' reward centers were not actually stimulating pleasure centers but a pathway leading from the base of the brain (the ventral tegmental area, VTA), through the hypothalamus and terminating in higher centers, including the then little known nucleus accumbens and the frontal cortex. It was further discovered that this tract, the midbrain dopamine pathway, was concerned almost exclusively with the transmission of the neurotransmitter dopamine from the cell bodies in the basal forebrain (the ventral tegmental area (VTA) is at the very bottom of the brain near where the spinal cord connects to other brain structures) to the higher level centers that controlled motivation and choice (the nucleus accumbens and frontal lobes). This gave rise to the idea that dopamine was the neurotransmitter that controlled the sensation of pleasure. It further implied that addictions were rooted not only in the processes of replacing, enhancing or otherwise changing the action of neurotransmitters at the synaptic junctions but that it represented a response to a possible lack of dopamine in these centers. From this perspective, the addict could be understood as dopamine deprived and therefore impaired in his capacity to enjoy the normal pleasures of life.

Continuing research, however, discovered that there were problems associated with the dopamine depletion hypothesis. First, if dopamine were the pleasure transmitter, how was it that addicts continued to seek drugs even after, as they complained, the drugs no longer gave them pleasure? Second, if dopamine and the Nucleus Accumbens were responsible for all reinforcement, how could you explain the observation that dopamine deficient animals and animals who had had their Accumbens Nuclei removed or ablated, still responded to natural rewards like sweet tastes and drinking when thirsty (Berridge & Robinson, 1998)?

By the standard rules of reinforcement theory (Ferster & Skinner, 1954), most addictions to substances of abuse and most behavioral addictions should disappear on their own as they become less and less rewarding. However, even though over time, addicts report lessened pleasure from the drugs or behaviors (decreased hedonic impact); they complain that they still want the drug. This has led researchers to focus not on the pleasure that drugs impart (hedonic impact) but on their ability to create craving or wanting (incentive salience). It is this factor, craving or wanting, that is mediated by the midbrain dopamine system (Robinson, 2004; Robinson & Berridge, 2001).

Incentive salience connects to neurophysiology through a series of experiments on single dopaminergic neurons and neural implants measuring the response of the neurons to various stimulus conditions. In general, researchers found that the midbrain dopamine system responds in very specific and predictable ways. First, it responds powerfully to novel rewards. Whenever rewards appear in an unexpected context, these neurons respond vigorously. Second, the brain seeks "the difference that makes a difference." If a stimulus fully predicts a reward, the neuronal response decreases. If a predicted reward fails to appear, the neural response decreases or disappears (This is one of the neural roots of habituation.). Third, if the stimulus predicts a reward that appears reliably but increases in value compared to other recent rewards, the neurons again increase the intensity of their response (Schultz, Dayan & Montague, 1997; Robinson & Berridge, 2001; Waelti, Dickenson, & Schultz, 2001; Robinson, 2004;

Tobler, Fiorillo & Schultz, 2005). These reactions determine whether a stimulus is more or less worthy of attention, whether it warrants immediate action or whether it can or has been automated and need not take up attention.

This research has established that both drug related and most normal motivations are rooted in the action of dopamine in the midbrain dopamine system. Furthermore, all drugs of abuse, whether directly or indirectly, create the cravings associated with addiction by stimulating the production of dopamine or preventing the reuptake or dismantling of dopamine in this area. It has often been reported that drugs of abuse hijack normal motivations by flooding the nucleus Accumbens and the brain more generally with dopamine (both intra-synaptic and extra-synaptic), thus granting the drug and its related behaviors, associations and circumstances increased significance and value (Goldstein & Volkow, 2002; Leshner, 2005).

The mechanism of addiction begins with pleasure, or hedonic impact. When the use of addictive substances is first learned or acquired, early motivations are dominated by the pleasures provided by the drug and the associations surrounding its use. This is mediated by endogenous opioids (neuro-modulators that are often associated with pleasure) in the core region of the nucleus accumbens. Inputs from the amygdala and the hippocampus label the stimulus as pleasurable or unpleasurable based on past experience or abstract knowledge (Baler & Volkow, 2006; Chambers, Taylor et al. 2003; Bechara, 2005). However, it is the incentive salience, the importance or perceived value of the drug or behavior, mediated by the flow of dopamine in the nucleus accumbens that explains the craving which is the main feature distinguishing addiction from abuse or dependence. Montague, Hyman and Cohen (2004) indicate that because drugs of abuse directly impact the ability of the nucleus accumbens to assign value or importance to a stimulus, they, along with their associated behaviors and circumstances are moved to the head of a salience hierarchy marking them as behaviors and rewards that tend to become preferable to all others (Hyman et al., 2006).

Once the habit is well established, the locus of control shifts from the nucleus accumbens in the ventral striatum (part of the basal ganglia in the center of the brain) to other centers in the dorsal striatum that mediate automatic behavior (Chambers et al. 2007; Feil et al., 2010).

Preference hierarchies are created by neurons that ascend from the nucleus accumbens to the orbito-frontal cortex (In the front of the brain, just above the eyes). A general pattern that appears in other parts of the brain (Craig, 2009), arranges hierarchies with the most valuable, salient stimuli nearest the center of the brain and those that are more abstract and less salient towards the frontal poles. Moreover, Davidson reports that positive and negative hierarchies are separated and represented in separate hemispheres. Approach-valenced or positive motivators are centered in the left orbito-frontal cortex and avoidance-valenced hierarchies in the right. Kringelbach's more recent meta-analysis suggests that positive reinforcers are centered towards the middle of the vMPFC and punishers that result in action, towards the sides. In any event, positive and negative hierarchies appear to be separated from one another (Davidson, 1993; Kringelbach, 2005).

According to Kringelbach, one of the important things that happens in the orbito-frontal cortex is that sensory inputs regarding experiences of various kinds are integrated into a multi-sensory representation of the experience. The richness of that sensory experience provides some of the information by which the stimulus is accorded its place in the hierarchy. When a more salient stimulus, one that seems to be more important, more fully represented in perceptual space, more crucial to survival, or more fully associated with a chemical rush arises in experience, those experiences that have not stimulated so strong a response in the nucleus accumbens are devalued and move further down in the hierarchy. Because they tend to directly stimulate, imitate or otherwise enhance the dopamine response in the motivational circuits, drugs of abuse tend to overpower most other things in the hierarchy. As a result, they tend to become the most important things in a context or in a person's life.

At a chemical level, drugs directly affect the perceived importance of the drugs themselves, the persons related to the drugs, the places where the drugs are taken and the results that they provide. They can do this because:

- Drugs directly impact the centers of motivation at a chemical level;
- The impact of drugs is relatively immediate, they are therefore perceived as better answers to life's problems than other rewards;

Both of these factors impact the way that the motivation system responds to drug related stimuli—including the drugs.

As a result, drugs may be accorded greater importance (incentive salience) than other stimuli, actions, behaviors, people, places and things and so move to the head of the preference hierarchy.

When we look back to the morphine drinking rats (Alexander et al., 1981), we see that rats with no options preferred morphine-laced water, but those with more important things to do, passed it by. On the level of *rat life*, sex, activity, fighting and socializing are far more important than drugs.

As we look back on the GIs who became addicted to heroin in Vietnam (Robins, Davis & Nurco, 1974), we find that heroin may have provided welcome relief from the stress of war. It was an immediate, salient answer to the reality of death and chaos. However, when they returned home, those with more salient roles, identities and possibilities awaiting them, in the absence of the stress of war were able to walk away from the drugs into far more important things, like love, marriage, education and careers. For those who returned to stress or for whom there were no strongly valued directions or roles, drugs continued as the most salient answer to the problems at hand. For others, who may have gone to the war already addicted, their experience may have provided a role and an identity that was sufficiently strong to out frame the lure of addiction (Peale, Brodsky & Arnold, 1992).

As we remember that the actions of the motivational system make the substance or the specific stimulus object more important, and that they enhance the salience of the place or circumstance, we can perhaps understand that when pain is the context for the use of "addictive substances," then the release of pain will also signal the end of the importance of the drugs.

Here is a partial answer: For persons who become addicted in special circumstances, the end of the circumstance may signal the end of the addiction. That circumstance may be external, like small cages and limited activity, or the stress of war. It may also be internal like stress or pain. If this is so, there may be other internal contexts that can serve to out frame addictions more generally.

Defects in Inhibition and Behavioral Monitoring

To this point we have described the incentive side of the addiction model, the part that deals most centrally with why drugs can positively overpower other motivations and behaviors and in turn, how more salient motivations may be able to overcome addictions. On the other side, as already mentioned in our brief discussion of Bechara and Cloninger, there are deficits in behavioral monitoring and response inhibition that contribute to addiction and may set the stage for the development of addiction.

Bechara (2005) proposes that addiction is characterized by a defect in the opponent process system that guides behavior. While the incentive system already described is driven by impulses that mimic survival responses, the inhibitory functions, centered in the frontal cortex, are ordered by learned values and personal experience. Prior experiences whether real or imagined—based on warnings and learned rules—are typically evoked by current events

It is now accepted by many researchers that memory serves a predictive function and serves to prime perspective and understanding (Nadel, Hupbach et al, 2012; Kroes & Fernandez, 2012, Williams & Bargh, 2008). Because current experiences become associated as permanent parts of preexisting memory structures (think of categories and mind maps) more efficiently than when learned without some pre-existing context, we are often controlled by our previous experiences or the currently active perceptual set (Bargh, 1997; Morris, 2006; Tse, Langston, Bethus, Wood, Witter, & Morris, 2008; Williams & Bargh, 2008). Bechara indicates that this mechanism is mediated through the inhibitory functions of the vMPFC which relies on an intact dorso-lateral prefrontal cortex (DLPFC) to perform its function. He notes that one of the problems accompanying addiction is characterized by defects in the DLPFC. Among other things, these deficits make it difficult to focus attention, especially on long term goals. This is reflected in research that shows an increased tendency for persons addicted to any of a number of substances to express a preference for lesser immediate rewards as opposed to larger but more distant rewards. This was already discussed with regard to Bechara's studies on the Iowa Gambling test. Bickel and colleagues, have shown that addiction spectrum problems bias decision making of all kinds in favor of more immediate rewards. (Bechara 2005, Bickel, Kowa et al., 2006; Bickel, Miller et al. 2007; Bickel Yi et al., 2008).

As drug related behaviors are promoted to the head of the salience hierarchy, the increased presence of extracellular dopamine reduces the number of dopamine (d2) receptors in circuits that are not related to drugs or drug contexts. The major brain circuits that suffer from these depletions are primarily concerned with the evaluation of behavioral outcomes and the inhibition of unwanted behaviors. These include the following circuits: those arising in the dorsolateral prefrontal cortex (DLPFC) which choose behaviors and percepts, focus attention, plan for their execution and allow for flexible transition from one focus or outcome to another; those originating in the anterior cingulate cortex (ACC) which evaluate behavior, predict errors and override inappropriate behaviors; and those originating in the right ventromedial and orbitofrontal cortices (rVMPFC, rOFC) which modulate automatic behaviors, limit impulsivity and socially inappropriate behaviors, and provide contextually appropriate evaluative feedback. Addictions dominate the user's life through this combination: the increased valuation of the addiction related substances and contexts, and the decrease in evaluative capacities centered in the DLPFC, ACC, VMPFC and OFC (Bechara, 2005;

Bechara & Damasio. 2002; Bechara, et al 2000; Bechara, Damasio, et al 1999; Bechara, et al, 2002; Bechara, & van der Kooy, 1985; Canales, 2005; Craig, 2009; Davidson, 1993; Diekhof, et al, 2008; Feil, et al 2010; Goldstein & Volkow, 2002; Kringelbach, 2005; Volkow, et al., 2002).

Dissociation of behavioral networks

Beyond the above considerations, addictions are characterized by one more complicating factor. As they bias neural systems to prefer addiction-related situations and associations, they give rise to associative networks that are increasingly centered on the people places and things associated with the addictive substance or behavior. This is similar to the literature of extinction and inhibition in classical conditioning that has also shown that one set of associations tends to dominate a response network while excluding the influence of other possible associations. State dependent memories reflect the almost absolute dissociation between contextually bound behavioral systems. Chambers has suggested that behavioral networks exist independently of one another and that the evocation of one network can effectively render another temporarily inaccessible. Craig (2009) has argued for the functional division of forebrain structures into those subserving sympathetic and parasympathetic activities as major subdivisions of emotional control. (Bouton, 1994; Bouton & Moody, 2004; Canales, 2005; Chambers, et al., 2007; Craig, 2002, 2009; Rescorla, 1988; Rossi, 1986; Rossi & Cheek, 1988).

This is relevant to addiction in that a drugged state, like alcohol intoxication, is often inaccessible to normal waking contexts and often leads to behaviors that are segregated into specific drug related contexts. This is intuitively recognized by the popular idea that people are different when drunk or stoned—"He isn't like that; it was the alcohol speaking."

The classical test for addiction in animal studies (so called pre-clinical studies) is a place preference association. That is, in both human and animal studies, the addicted subjects prefer the places and circumstances where the get or have gotten high. In the context of addictions, the assertion of a novel context, not associated with addictive response systems, may be crucial to recovery. The relevance of the findings of the rat park study, the spontaneous recovery of the Viet Nam Veterans and the low level of addiction among those who use opiates for pain remediation is that each of these circumstances are represented by separate neural networks that exist independently of one another. When one is salient, the other is relatively inaccessible. The addicted life may run parallel and separate from the potentials that could provide an answer (Alexander, Beyerstein, Hadaway, & Coambs, 1981; Colleau, 1998; Peele, Brodsky & Arnold, 1992; Robins, Davis, & Nurco, 1974).

Target Problems

The preceding evidence suggests that there are three issues that need to be dealt with in answering the problem of addiction. The first is the overwhelming incentive salience of the addictive behaviour. The second is restoring the active capacity of attentional and evaluative mechanisms in the orbito-frontal cortex, the ventromedial prefrontal cortex, the anterior cingulate cortex and the dorso-lateral prefrontal cortex. The third problem is ensuring that alternate behavioural networks are in place that are independent of the drug centred behavioral networks and that they are sufficiently self-reinforcing and salient that they will maintain sober behaviour over time.

Chapter Five

Dimensions of Motivation:
A motivational primer

Motivation is a crucial factor in the treatment of addiction. Modern neuroscience suggests that the problem of addiction is precisely a problem of motivation and its chemical manipulation by substances of abuse. Yet our understanding of motivation and how to manipulate it positively continues to be informed by moralistic and common sense judgments rather than insights from a growing body of psychological and neurophysiological data (Baler & Volkow, 2006; Goldstein & Volkow, 2002; Shattuck, 1994).

Common sense notions of motivations often spring from the fundamental attribution error. We look at the behavior of others and assume that what appears logical to us must also be logical to them. When they fail to respond according to our understanding we see them as flawed, sinful or broken. Their failure to live up to our expectations must be due to some internal fault or trait. When, however, we find ourselves in the same situation, we have a perfectly reasonable excuse based on external pressures or influences beyond our control; our failings are held to be completely rational (Gilbert & Malone, 1995; Shattuck, 1994).

When we consider the motivations involved in stopping substance use, whether at the level of abuse or addiction, the common sense notion is that the consequences should speak for themselves: "If *I* were in that situation…" When our subject fails to respond according to our expectation they become *to us* hopeless addicts.

There are three salient points that such a perspective misses. The first is that, by definition, addiction and substance abuse are characterized by a loss of control. The second is that negative consequences are poor motivators, especially as regards addictive behaviors. The third is that motivations are always held in regard of specific objects, goals or outcomes (Baler & Volkow, 2006; Bechara, 2005; Chambers, Taylor et al., 2003; Gray, 2005, 2008; Inaba & Cohen, 2007; McKim, 2003).

We have already covered the first point. The second, however, bears further inspection. If we look back over the work of Bechara (2005), Cloninger (1987), Chambers et al. (2007), Feil et al. (2010), Reddish et al. (2008) and others, we immediately discover that in the advanced forms of substance use disorders, the neural structures involved in assessing negative consequences (the DLPFC, OFC, ACC) were either dysfunctional before substance abuse began or became dysfunctional in its wake. The idea of consequences means very little to someone deeply involved in an addictive spectrum problem.

In their 1999 book, *A Brief Guide to Brief Therapy*, Cade and O'Hanlon provide several pages of things that don't work to change people. Their list is relevant to multiple contexts but most especially to addiction spectrum disorders. They start with the following kinds of unsolicited guidance:

> ... lectures, advice (especially when given 'for your own good!'), nagging, hints, encouragement (Why don't you just try to....), begging, pleading, trying to justify your position, appeals to logic or to common sense, pamphlets/newspaper articles strategically left lying around, or read out loud, the silent, long-suffering 'look at how patiently and bravely I am not saying or noticing anything' approach, or an angry version of the same (these are often the most powerful 'lectures' of the lot), repeated and/or escalating punishments tend also not to work and often result in more of the same, or an escalation of, problem behaviors. (p.82)

They continue by pointing out how these already ineffective techniques become even more galling to their targets when they are framed from the moral high ground of a supposedly superior moral or intellectual position. The frame is often set as follows: "If you really loved me..." "Surely you could see that if you..." "Why can't you realize that...?" "Anyone with any sense..." "After all I've done..." "Look how ill/desperate/depressed I've made myself by worrying about..." "I love you because you behave as I want you to..." (p.82).

We forget that drugs, alcohol and other problem substances and behaviors are the go-to resources for the people who constitute this client base. When they become stressed, the problem behavior or substance is the answer, it is immediate, intuitive and it works every time. This is why it stands at the top of the preference hierarchy. So, when we apply pressure, 'tough love', get angry or insistent, our net effect is often to make things worse. They get to soothe themselves in the sure knowledge that their solution works.

Prochaska (1994) found that people are most likely to change when they have identified a positive outcome that is more important to them than the problem behavior. Once that outcome has been identified, people naturally begin to devalue the problem behavior. Motivation towards a salient outcome devalues less salient outcomes. It lies at the heart of addiction spectrum disorders and is a key to solving them.

If the primary problem in addiction is understood to be the motivation to use, incentive salience, then one of the most important places that we can look to find answers is in the possibility of creating alternate outcomes. The following information on motivation is aimed specifically at motivating those kinds of behaviors that can actively compete with the use and misuse of various mood altering substances and behaviors. It assumes that strong intrinsic motivations are often the key to turning away from problem behaviors, towards more productive, more meaningful and more life affirming directions.

MOTIVATIONS

Writing in 2008, Hulleman, Durik, Schweigert and Harackiewicz propose a model of motivation where tasks become desirable as they become achievable and interesting to the subject. The more interesting the task and the more confidence the actor experiences in her capacity to accomplish it, the more highly motivating the task becomes. Not only so, but if the task contributes towards outcomes that are already valued and the task is itself viewed to be interesting, this combination will result in increased interest and willingness to approach the task.

This agrees with what we know about the lessons of the rat parks, and the lessons that Peele et al. draw from both the rat parks and addicted GIs; if there is something that is more fundamentally meaningful, interesting or empowering, it can compete successfully with the problem behavior.

In 2008, Deci and Ryan reported that long term goals often fell into two broad categories. The first category included such things as money, fame and being attractive. These were labeled extrinsic motivators because they focused on external indicators of worth that relied on the judgments, or perceived judgments of other people. The second category, labeled intrinsic motivators, focused on things like personal growth, relationship building, and participating productively in community life. They found that intrinsic goals were more likely to be successful, conducive to mental health, and more likely to encourage follow through.

As early as 1975, William Nootz, in a review of then current research, indicated that, with regard to motivation for achievement of almost any kind, the single most important factor was whether the motivation was intrinsic or extrinsic. He found that intrinsic motivators were generally more fulfilling, were more likely to provide successful outcomes and in general were superior foundations for purposeful behavior. Extrinsic motivators often produced less than stellar results and often interfered with performance that had previously been successfully supported by intrinsic motivators.

From the perspective of Self Determination Theory (SDT), Deci and Ryan further classified motivators in terms of whether they were freely chosen or involved some threat or consequence: as autonomous or controlled motivations. Autonomous motivation was understood as including both truly intrinsic motivators (things wanted for their own sake) and well internalized extrinsic motivators. They found that autonomous motivations that satisfied basic needs for competence, autonomy and relatedness tended to promote intrapersonal well-being, goal satisfaction and focused activity with regard to outcomes. Their research has shown that the type of motivation is often much more important in determining success than the intensity of the motivation.

> Autonomous motivation involves behaving with a full sense of volition and choice, whereas controlled motivation involves behaving with the experience of pressure and demand toward specific outcomes that comes from forces perceived to be external to the self (p. 14).

These authors (Deci and Ryan, 2008) also reported that the impact of extrinsic rewards was often mediated by the spirit in which they were given. A great deal of literature had previously shown that extrinsic rewards could often spoil intrinsically motivated behavior. Paying someone for doing the thing that they loved would often be found to destroy the pleasure of doing it. But that effect was uneven and often varied across contexts. Deci and Ryan indicated that the social context of the external reward often determined how it affected otherwise intrinsic behavior. If the reward were given as feedback, in a spirit of cooperation, it often either had no impact on intrinsic motivation or added to the level of motivation. If, however, the reward or feedback came in a way that could be understood as control or criticism, it could negatively impact performance.

> … [Al]though tangible rewards have been found to undermine intrinsic motivation, if the interpersonal context is informational and supportive of people's autonomy, the rewards could have a positive effect (Ryan, Mims, & Koestner,

1983). In parallel fashion, if positive feedback is administered in a controlling context, it will tend to decrease (rather than increase) intrinsic motivation (2008, p.15).

Hulleman, et al. (2008) also reported that other researchers had found that tasks are perceived as more motivating if they possess some intrinsic value for the subject. Tasks possessing intrinsic value are likely to be perceived as enjoyable and fun. When tasks are perceived as having utility they may be understood as being not just useful for my immediate purposes but are relevant to other aspects of life (p.398). Consider the complaint of the student who, faced with a class or subject that is not interesting asks the question: "How is this relevant to my life and the things I'm planning to do?"

Koestner (2008) reports that when people set goals and are able to follow them through, those goals are most often autonomous. Their pursuit is ensured by the creation of implementation decisions—clear plans for goal pursuit that "…facilitate retrieval of goal intentions in memory, heighten accessibility of environmental cues for goal completion, and reduce the number of interruptions while one is in goal pursuit" (pp 63-64). To a large extent, goal pursuit involves a process of self-remembering; people who remain focused on their own reasons for choosing an outcome do better than those who work to overcome external influence (Koestner, 2008).

According to Baumeister and Heatherton (1996) an important aspect of setting and attaining goals is self-regulation. Closely related to motivation, self-regulation speaks of the way we keep on track in attaining those goals. There are three elements that affect self-regulation with regard to outcomes and goals. The first is the presence of an end-state, that is: a goal, standard, ideal or something else that signals completion. These conceptions must be clear and consistent.

The second is self-monitoring—remaining aware of the outcome, what needs to be done, and whether or not it is being accomplished. The third is capacity or efficacy—whether it is within their capacity and under their control.

Maintaining control over goal oriented behavior often requires us to override a behavior that has already begun. The capacity to override a behavior depends upon the relative strength of the behaviors involved. Note that this suggests that a more valued goal is more likely to dominate a less-valued goal. According to these authors, there is a limited amount of energy available for self-regulation and it must be used appropriately. Baumeister and Heatherton note that self-regulatory strength varies from person to person and from situation to situation. Criminals often lack in self-regulatory strength across life domains. People who believe that washing their hands 25 times is crucial to their safety, may find that behavior hard to overcome. People who suddenly discover their life's purpose, experience a spiritual awakening or truly and deeply fall in love, may find it easy to leave other things behind.

The same authors indicate that because self-regulatory strength is limited, trying to do too much will make even normally easy tasks much more difficult. This implies that if outcomes or directions are treated as independent events, they will compete one with another. An implicit counter to this notion of split self-regulatory energy is the possibility that very deep motivations may entrain the energies of lesser outcomes. When they all move congruently in support of a unified goal, any energy investment is understood as a contribution to the whole. This is similar to the idea from Hulleman and colleagues that when an outcome accords with or supports some other deeply desired goal, it becomes much more motivating.

Baumeister and Heatherton (1996) indicate that managing attention is the most effective way of managing behavior. It works hand in hand with the observation that the earlier in a behavioral sequence an intervention occurs, the more effective it is likely to be. They also point to transcendence as an important facet of attention control.

Transcendence means turning the attention towards larger or more important future outcomes and so reframing the more immediate stimulus in terms of the larger goal. Failure of self-regulation is failure of transcendence. Failures in transcendence involve a re-emergence of the importance of the immediate. The influence of others and the loss of individuation often provide the same effect. It may be important to note that transcendence as imagined by Baumeister and Heatherton is parallel to the Strong principle of change identified by Prochaska, Norcross and DiClemete (1994a, 1994b; Prochaska, 1994). In both cases a more compelling future outcome draws attention to itself and devalues competing end states

This highlights an essential facet of self-regulation that many of the authors noted here seem to miss in their insistence that that this energy is limited. Baumeister and Heatherton indicate that if we must fight too many battles for focus and the control of options, we may deplete our reserves of self-regulatory energy and end in failure. These authors and others fail to consider the possibility that motivations may be constructed synergistically so that each supports the other. In this case, energy expended on any one facet of the program benefits the whole.

This type of structure is implicit in Jung's idea of the Self and the process of Individuation. He notes, as do his later interpreters (Gray, 1996, Hillman, 1977), that every complex, every action and impulse implicitly reflects the draw of the Self. In the non-individuating person, the complexes, urges, outcomes or directions, compete with one another for limited resources of psychic energy. When, however, tasks are marshaled towards the conscious realization of the Self, the effort focused on any one aspect serves the whole. Likewise, in Maslow's idea of self-actualization, the individual options narrow so that energy is only expended on those that subserve the greater goal of self-actualization (Gray, 1996, 2011; Maslow, 1970).

In such motivational circumstances the benefits of the systems principle of wholeness and emergent properties move the actor from the realm of limited, discrete portions of self-regulatory energy to a condition where any exercise of that energy reinforces the entire system. In line with Jungian and Maslowian principles we understand that the deeper and the more fundamental the outcome is, the more likely it is to participate in this level of systems organization (Gray, 1996, 2011).

From this brief review we come to understand that there are some basic criteria for creating or defining motivating outcomes. They include:

1. The outcome should be intrinsic; it should be valued and interesting for its own sake
2. The outcome should enhance autonomy and choice; it should increase—not decrease—options
3. The outcome should be well-defined. We should have a clear sense of what we want and how we will know that we have gotten it
4. The outcome should be under our personal control; not only as an option that we might choose, but as

something that we are capable of; something consistent with our self-efficacy beliefs
5. The more deeply felt the outcome, the more it is an expression of deep identity or calling, the more motivating and unifying it will be
6. The deeper and the more powerfully motivating the outcome is, the more capable it will be of sustaining the capacity to transcend competing goals.

(Andreas, & Andreas, 1989; Bodenhamer & Hall, 1988; Cade & O'Hanlon, 1993; Dilts & Delozier, 2000; Gray, 2011a; Linden & Perutz, 1998)

In this light, we can briefly mention the third point from the beginning of the chapter. Motivations are always held in regard of specific objects, goals or outcomes. Whether the outcome is the single object of effort or the broader goal of personal growth, insofar as it can be specified, it can become the center of meaningful effort.

There are two other broad motivational structures that should probably be mentioned here, Quantum Change and Flow states:

In the mid-1990s, William Miller and Janet C'deBaca placed an ad in an Albuquerque, NM newspaper inviting people who had experienced sudden and dramatic life changes to contact them for interviews about their experiences. After a flood of applications, Miller and C'deBaca chose and interviewed 55 subjects. For some, the changes came at the lowest point in their lives. For others, the changes came in a period characterized by a stagnant normalcy. All of them reported sudden, dramatic, positive and permanent life changes. The changes ranged from complete transformations of life to the loss of a persistent addiction or other pervasive problems.

There were two kinds of changes. One was characterized as mystical and involved the perception of some mystical other. It involved intense feelings of oneness, love and forgiveness and was often accompanied by ecstatic states. The other was characterized

> ... as centering on insight, something lying more within the conceptual world of psychotherapy. These stories lack most of the classic components of mystical experience save one: the noetic element of sudden realization or knowing. Such insights are distinctively different from the "a-ha" insights of ordinary experience. These awakenings break upon the person with great and sudden force, and in the moment of seeing, the person recognizes them for authentic truth (or Truth). Their effect tends to be a reorganization of one's perceptions of self and reality, usually accompanied by intense emotion and a cathartic, even ecstatic, sense of relief and release. (Miller, 2004, p.457)

Miller noted further that many of these changers experienced complete release from long standing addictions and dependencies but, unlike typical recovery narratives, the quantum changers were not obsessed with the fear of relapse. They had apparently taken on the identity of a person who does not drink or have that problem. In effect, the problem had become irrelevant to their current life. Miller notes, "The person does not merely change behavior and stop drinking, but truly becomes—is transformed into—a nondrinker" (Miller, 2004, p. 456). In general, their priorities were radically changed and it seemed as if they had experienced Maslowian self-actualization at high speed.

Miller describes a particularly striking reordering of priorities in relation to hierarchies of values. He indicates:

Another major change that quantum changers reported was in their values and priorities. Looking back at their core values before their experience, men reported that their top priorities had been wealth, adventure, achievement, pleasure, and being respected; women said that family, independence, career, fitting in, and attractiveness had been most important. Both reported an abrupt and enduring shift in their most central values. After their quantum-change experience, men ranked spirituality, personal peace, family, God's will, and honesty most highly; women valued growth, self-esteem, spirituality, happiness, and generosity. They were no longer possessed by their possessions. Often, characteristics that had been valued least became most important, and those that had ranked as highest priorities fell to the bottom. Spirituality, though not necessarily religion, became central for many. Relationships were changed, too. Quantum changers often seemed to lose their tolerance for superficial relationships. They wanted fewer and closer friendships. Some experienced sudden healing of and release from enmeshment or abuse they had experienced in childhood. Others found the courage to leave abusive relationships. For some, family and intimate relationships became more meaningful and peaceful. (Miller, 2004, p. 47)

Another category of motivational structure is Czikszentmihalyi's idea of flow. Flow is that state where a person is optimally committed to a task: the individual merges with the task, ego consciousness seems to disappear, time goes away and the engagement is perceived as positive and empowering. According to Czikszentmihalyi, entire lives can be lived out in this state.

The essential elements of flow are as follows:

1. There is a well-defined task with a clearly stated outcome
2. The performer is aware that he has skills that are adequate to the task but the task always draws him on towards deeper engagement as what may have been a goal recedes just beyond reach
3. The performer is acutely aware of the demands of the situation and his own capacity to respond in an appropriate manner
4. The focus of the performer is limited to the task and task relevant variables Distracters recede from consciousness and the task itself becomes both means and end. The task becomes autotelic or self-reinforcing
5. There is a merging of action and awareness in an ongoing and absolute focus on the task at hand
6. The performer forgets himself as he becomes more aware of the actions and perceptions that draw him more intimately into union with the task.
(Czikszentmihalyi, 1991; Czikszentmihalyi and Czikszentmihalyi,1988)

It is not difficult to observe how the crucial elements of well-formed, intrinsic motivators are represented in the flow state. There is a positive goal. It is under the performer's control and he knows how to do it. It is specifiable in terms of sensory experience. It is interesting and intrinsically motivating. These are criteria which, when awakened, will more than out frame any addictive state. In fact they constitute in a precise manner what Glaser (1985) has called 'positive addictions'.

As noted above, Czikszentmihalyi and Czikszentmihalyi (1988) suggested that whole lives can be lived out in a flow state.

Ultimately, this is life lived out as a spiritual path. It is a life filled with meaning and personally relevant direction. It is a life lived under the influence of what James Hillman (1996) describes as a calling and Maslow (1971) terms the path of self-actualization.

MOTIVATION AND ADDICTION

From the following we can glean some important information about motivation as it impacts addiction spectrum disorders.

1. Positive motivations and addictions are mirror images of the same neurological structures: we approve of one and disapprove of the other.
2. Because motivations are arranged in hierarchies, the more important the outcome, the more likely it will be pursued.
3. There are natural states that naturally and necessarily overpower addictions and other motivations; these are: powerful intrinsic motivators, quantum change, and lives lived out in flow.
4. Another, important element, related to motivation, through its effect on the rest of the brain, is the strengthening of the mechanisms of choice, self-reflection and inhibition, located in the frontal cortex.

Chapter Six

Hierarchies and preferences

When we consider that the brain sets up hierarchies of values, it is important to realize that those hierarchies are both dynamic and, to a large extent, context dependent. There are things that you want or enjoy thinking about in some contexts which would—or might be—unthinkable in another. For many people, thinking about sex in a cathedral is not possible. So, place and context become very important.

When we discuss addiction spectrum disorders, we can perhaps all remember a time when our friend the smoker forgot to have a cigarette and our friend the drinker had no need for a beer. These situations often identify meaning contexts where the hierarchy is changed by the current environmental context. For the moment, the problem behavior is out framed by more salient options.

In the studies of the morphine drinking rats, they found that in the rat parks there were behaviors and opportunities to behave that made the consumption of morphine less important. In those contexts morphine consumption dropped significantly. For rats, these behaviors included sex, exploration, fighting and other social behaviors.

In the heroin-addicted GIs, context may have been the crucial determinant of the continuation of addictive behavior. When confronted with the stress of war, especially an unpopular war where it was often impossible to tell friend from enemy, internal and external stressors—and the very real possibility that there was no future—provided a context that made the use of heroin very probable. Its immediate utility lay in its capacity to ease stress and make the horrors of war disappear. When the distress of war was eliminated, when the hope of a future was restored, the heroin addictions disappeared,

Gray (2011) reports that the intensity of addiction spectrum disorders increases as the use of the problem behavior generalizes from one context to another. The more available the behavior, and the larger the number of contexts in which it is available, the more problematic the behavior becomes.

By this logic, we can understand why, independent of their chemical properties, tobacco and alcohol are so pervasive and so difficult to quit. Both are sanctioned for use in multiple contexts. On some level, both have been part of coming-of-age rituals and marks of either social inclusion or rebellion. Alcohol is integrated in the most significant rituals of Judeo-Christian religious practice.

Beyond tobacco's declining popularity in the U.S., it is a drug that is easily integrated into every facet of life. One informant indicated that, when he stopped smoking after a period of heavy daily use, he found that after the urges had left there were motor patterns, patterns of behavior where he expected to have a cigarette to complete an action or thought. In the past, when he wanted to start a conversation, he could ask for a light or ask if his companion minded if he smoked. When he was stuck for a reply or could not immediately answer a question, he could pause to think as he lit up. When he needed to take a break he could go out for a smoke. Cigarettes had become an ever-present aid. This matches Gregory Bateson's observation that addictions create behavioral circuits and that some of the difficulties of stopping addictive behaviors are caused by the perseveration of those motor habits. Note that these are not urges, per se, they are patterns of behavior that have no meaning without the problem substance or activity (Bateson, 1972).

Drugs and alcohol often work the same way. They may begin as social lubricants or as part of an acceptance ritual. Luigi Zoja (1990) points to the initiatic impact of drug and alcohol abuse. When someone is offered the opportunity to join the fellowship of drug users, there are implicit initiatic passages. 1. On the most basic level, there is an offer of acceptance and belonging. No matter that it may be to an aberrant group, it is a welcome. 2. The welcoming group is taking a risk by inviting the neophyte into their midst. Because the drugs are illegal either by statute or because the neophyte is underage for their consumption, there is a risk in the invitation. The willingness to risk the safety of the group or host means that the newcomer is valued sufficiently for the other members to take a risk. 3. The drug or other substance provides a direct experience of an altered state of consciousness which, subjectively and chemically, raises its importance. If the new initiate is insufficiently grounded in the hope of a positive future, if they are not invested in their own lives, the above factors make the experience much more valuable than it might be otherwise. Even if they are relatively well invested, the chemical impact of the substance multiplies its own perceived value.

One of the most important sources of human motivation is social reinforcement. Much of modern behavioral research into drug treatment focuses on social reinforcement. The Twelve-Step fellowships depend (implicitly, if not explicitly) on the power of social bonds and group support to help the transition from substance use to abstinence. Because of this, it is important to realize that the use of mind altering substances almost always arises in a social context.

Gray (Private communication, 2008) reports that during more than 25 years of working with state and federal offenders, many of them addicted or dependent upon a variety of substances, he never encountered a case where the initiation and maintenance of such behaviors did not have their roots in a social context or that were not maintained, at least in part, by social reinforcements. He recounts the story of a young woman who was introduced to heroin by her friends. Her first several trials made her violently ill. Her friends, however, encouraged her and told her that it takes a while to get used to it. With their help, she persevered and eventually found that the nausea was replaced by much more interesting sensations.

In another anecdote, Gray reports how in the 1960s novices being introduced to the use of marijuana would often experience paranoia, headaches and other problems. With the aid of more experienced users, they learned to banish those problems with stereo headphones, visual distractions, a little wine, some mild tranquilizers and chocolate chip cookies.

In order to understand the basic ordering of these hierarchies we would do well to return to the basic responses of the dopamine neuron where, it seems, the root patterns of motivation are encoded.

Single dopamine neurons function in the following fashion:

1. They fire robustly in response to novel rewards. Whenever rewards appear in an unexpected context, these neurons respond vigorously.
2. As the brain seeks "the difference that makes a difference," if a stimulus fully predicts a reward, the neuronal response decreases (This is called habituation). If a predicted reward fails to appear, the neural response decreases or disappears.
3. If the stimulus predicts a reward that appears reliably but increases in value relative to other recent rewards, the neurons again increase the intensity of their response (Schultz, Dayan and Montague, 1997; Robinson and Berridge, 2001; Waelti, Dickenson, and Schultz, 2001; Robinson, 2004; Tobler, Fiorillo and Schultz, 2005).

From this list we can begin to understand how a stimulus comes to be accorded high levels of incentive salience. The first criterion seems to be whether the organism got something that they didn't expect and that thing was in some way good or useful. Animals respond in order to have the opportunity to experience something new. Humans are often consumed with neophilia and the need for stimulation. People get bored. One of the important things that drugs have to offer is novelty. To effectively challenge them, our interventions must also include something more than just information[4] (Daly, Mercer & Carpenter, 2002; Diekhof, et al 2008; NIDA, 2002;

[4] This may be why educational treatments do not work for addiction spectrum disorders. It's more of the same and is not perceived as novel or interesting.

Robinson, 2004; Robinson & Berridge, 2001; Schultz, et al 1997; Tobler, et al 2005; Waelti, et al 2001).

If we were to structure an intervention that would take advantage of this neural characteristic, we would want it to be surprising. When William Miller (2004) described the factors that seemed to be an important part of quantum change, where peoples' lives were transformed in very short space of time, he noted that one of the important things that happened was that they were surprised by the revelation or the experience that they had. According to Colin Wilson (2002), this is a facet of what Abraham Maslow (1971) called peak experiences and G. K. Chesterton (1908) called "absurd good news." On a physiological level, we are primed to respond to surprises and mark them out in our physiology.

Because, according to principle number two, behaviors habituate, an intervention must do something other than just repeat itself. There must be some variability. If a stimulus or a rewarding behavior fails to provide new kinds of experience, it either becomes a simple, unthinking, conditioned response, or it disappears. When, however, a stimulus varies in its intensity and predictability, it entrains behavior. Ferster and Skinner (1954) understood this in terms of schedules of reinforcement. If a behavior is established by reinforcing it every time, then, when you stop rewarding it, it just disappears—it is extinguished. When a stimulus becomes relatively unpredictable—sometimes it works, sometimes it doesn't, sometimes it works better, sometimes not so well—it tends to become highly resistant to extinction.

In the development of addictions, the quality of the drugs, the availability of the drugs and the consistency of the environment can all vary. Because of this variability, the drug attains more salience and becomes more resistant to extinction. In the same manner, a pattern of abstinence, moderation or a program of recovery must make room for novelty, for "absurd good news." One of the important ways that this is done is by conceiving of recovery or life beyond the problem as a path or journey.

The third method whereby addictive spectrum problems arise is that as the novice learns how to use and appreciate the state, the state—for a certain time—gets stronger. Further, because responses to problem states and behaviors have a tendency to habituate, the reinforcement pattern tends to follow a series of plateaus and peaks as the user learns to titrate the dose. In general, as the dose-impact decreases, he or she finds that more works better and the pattern is reinforced.

In planning a strategy that will successfully reframe the problem state or behavior, we need to ensure that there are places for growth, places where the value of the new behaviors are enhanced appreciably. In short, *we must build the experience of hope.*

On a behavioral level, Austin and Vancouver, in their 1996 review of mechanisms of motivation and goal structures, indicate that behaviors, goals or outcomes are accorded higher priorities in salience hierarchies when they meet the following criteria:

- The behavior can be used as an integral part of different behavioral sequences ("I always have a drink before I go out, just to loosen up." "Whenever I have to face John's mother, I have a drink."). In the language of behavioral science we would say that the behavior is present in multiple schemas.

- It is found to be useful or available in multiple contexts (Cigarettes and alcohol become powerfully addictive because they are so well integrated into the contexts of everyday life.).
- A behavior becomes important when it seems to represent an easy answer, the path of least resistance. Drugs and behavioral problems work quickly and effectively to remove the stressors of the moment. They are easy, if impermanent, answers. In effect, the short term utility of the behavior and its generalization into multiple contexts tells the brain, "This is important!" (Gray, 2005).

If we were to apply these criteria to a strategy for addictive spectrum problems we may need to think about our intervention in a manner that focus not on the problem, but on alternative answers.

The first criterion indicates that our answer should have utility in multiple situations and it should be an important part of the changer's response system in those contexts. An identity or a deeply held belief might serve the purpose, as might a continuing means for adapting to stress or challenge. Ideally, it should be as a much a part of the changer's identity as the problem may have been. It could be something as simple as an easily accessible conditioned or anchored state, or confidence in personal capacities.

The second criterion emphasizes that it must be something that crosses conceptual and behavioral boundaries. When a man or woman falls in love in such a way that that relationship opens into a life path, that can be just such a resource. When people have had profound spiritual experiences that redefine who they are and their relationship to the present and the future, those experiences can often provide this kind of resource. If someone could simply change their frame of mind wherever they might be, that would also serve the purpose.

The tool should be easy, natural and intuitive. It should not be intellectual; it should be, as Robert Dilts likes to say, "… in the bones."

Another element affecting the hierarchical organization of preferences, beliefs and behaviors is found in the NLP meta-programs. NLP recognizes that people have preferences in the manner in which they encounter the world that serve as deep filters on their perceptions and preferences. These preferences were first described by Leslie (Cameron Bandler) Lebeau in the late 1970s. Lebeau originally identified 70 meta-programs. Later researchers found that many of these could be collapsed as they represented variants of larger categories. Later research by Bodenhamer and Hall has sought to re-expand the list (Bodenhamer & Hall, 1997; Charvet, 1997; Dilts & Delozier, 2000).

Meta-programs organize other thought processes. In general, they provide the broader context that might define why, in identical situations, attending to the same kind of stimuli, two people might respond very differently. Robert Dilts and Judith Delozier give the example of two people who make decisions based on seeing a series of objects, perceiving a feeling about those objects and making a decision based on those feelings. One subject indicates that upon seeing the examples, she feels better about one than any of the others. That leads her to make her choice. The other subject sees the objects, but her feeling response is overwhelming and she cannot make a decision. These two very different responses are governed by different means of dealing with the data of experience: meta-programs. So, meta-programs stand a logical level above the conditioned responses that create the feelings about the things that we encounter and they integrate those meanings into a frame of personal relevance. (Charvet, 1997; Dilts & Delozier, 2000)

Some of the more basic ways that people sort information and behavior using meta-programs include the following:

- Approach to the problem. Towards positive outcomes or away from negative consequences
- Time frames: Short term or long term and whether their orientation is towards the past present or future
- Chunk Size: Do they prefer generalities or details?
- Locus of control: internal –introverted, or external—extraverted
- Mode of comparison: Matching—finding similarities and uniformities or mismatching—finding differences and potential problems
- The approach to problem solutions: task oriented (whether by options or procedures) or oriented to relationships and their focus (self, other or communal)
- Thinking styles or channels: vision, action, logic or emotion—roughly equivalent to Jung's perceptual styles
- The preferred informational focus: People, places, things, information, procedures (Charvet, 1997; Dilts & Delozier, 2000)

These elements should be considered as part of the hierarchical structure of experience and may contribute significantly to how a person defines their response to the world. For example a person who has an external frame of reference may need the permission of an authority figure before they are ready to make a significant change.

During the late 1950s and into the early 1960s, my father was a heavy smoker. At that time, information about the ill effects of smoking was just coming to public notice, but most people still smoked. One day, after a physical examination, he was told by his physician that if he didn't stop smoking, he would die. As his external frame of reference and orientation towards authority made the word of an expert extremely valuable, he immediately took action and, in less than a month, he stopped totally. Other people, for whom the authority relations would be less valuable, might not respond with such fervor. For many people permissions and expert opinions are insufficient grounds for change; for others, they are.

There is a striking statistic that we would do well to consider. The limbic system, the seat of emotions, is the place where we experience the world on the level of what used to be called raw instinct. It *is* the seat of the emotions. Our culture, however, is very much focused on cortical capacities; thought, usually in the form of words. It might be said that one of the failings of our culture is the tendency to attribute to people more rationality than is reasonable. What we need to remember, however, is that the number of neurons that arise from the limbic system to the cortex outnumber projections form the cortex to the limbic system by a factor of many thousands to one. Recent research indicates that the amygdala, the heart of emotional responding, is the most richly enervated locus in the brain. As a matter of practical truth, it is the center of brain function (Damasio, 1999; Pesoa, 2008).

What this tells us is that words alone will not change people unless those words awaken feelings. Cognitive rules work over the long term as they come to awaken the feelings and meanings that make them relevant. Until they are well practiced and filled with personal meaning, they are only role-plays.

DILTS' NEUROLOGICAL LEVELS

In 1990, Robert Dilts published *Changing Belief Systems with NLP*. In that book, he set forth a system of neurological levels which explained the integration of various levels of belief and motivation. The levels were presented with regard to the structure and manipulation of beliefs. He also argued that these levels represented a hierarchy of neural involvement and complexity. The levels and their associated motivational frames (from broadest to narrowest) were as follows:

- Spirit or strategic vision - What is my intention or purpose for this? What does this mean?
- Identity - Who am I?
- Belief and values - What are my beliefs? What is preferable in this situation? What is the best answer for this problem?
- Capability - What am I capable of doing? This includes maps, strategies and the capacity to generalize.
- Behavior - What am I able to do?
- Environment - In what context does this behavior occur? What are the external constraints?

Presented by Dilts as rooted in Bateson's (1975) adaptation of Bertrand Russell's logical types, the model has often been criticized as being inconsistent with both Bateson and Russell (Bostic St. Clair & Grinder, 2002). Admitting that the structure has serious problems when held up to exacting philosophical criteria, we can understand its utility if we apply it as one of several possible orderings of salience hierarchies. As such, it helps to clarify some of the relations between motivations and actions.

Dilts' levels represent a systems theoretical model of the levels of control for various kinds of behaviors and perceptions. This implies that the structure and function of each behavior or perception is preserved at each level of function, but as each level is incorporated into higher levels, its meaning within the whole changes. From a systems theoretical perspective, change can occur from the bottom up or from the top down. From the bottom up, assemblies of systems or behaviors come to the point where their interrelations reach a level of complexity that redefines all of their functions in terms of a larger whole. This is an emergent property. From the top down, we understand that the higher, more integrative levels determine the meaning and purpose of the lower levels (Gray, 1996).

In ascending order, Dilts' levels represent stages of increasing complexity that emerge from the interactions of simpler behavioral systems. In descending order they represent control structures that inform or modify the meaning and behavioral salience of the individual behaviors and perceptions below them. It appears that the system works most appropriately as a means of understanding motivation and preference (behavioral salience—the likelihood of behavioral expression and incentive salience—the level to which one will work to achieve an outcome), as opposed to any other psychic element. On the purest level they appear to be levels of subjective organization.

Feil et al. (2010) describe just such an organization in their description of the assembly of smaller behavioral elements into larger schemas in the ventral striatum. They indicate that motivation on its most basic level is concerned with the assembly of successive acts towards the accomplishment of a larger goal that is set by higher control mechanisms.

It is important to realize that, as a matter of practical application, Dilts' levels represent a recursive system, that is, a system that repeats itself on multiple levels. The first three levels may be executed in a fully unconscious manner or they may represent levels of increasing consciousness and choice. When they operate to reveal increasing levels of consciousness, they implicitly incorporate the same kinds of transformations on an unconscious level.

At the most basic level of the hierarchy, there are stimulus response interactions which are automatic and are controlled by environmental variables. These give rise to reflex actions and mood changes. They are subsumed into larger behavioral units which are relatively more conscious and are subject to choice. Dilts calls these larger elements behaviors. Behaviors tend to be more conscious and can be related to operant behaviors as opposed to the more Pavlovian, stimulus bound behaviors at the environmental level.

The awareness of behaviors and their possible application to multiple contexts gives rise to the perception of capabilities. These have also been referred to as efficacy experiences by Bandura and others. Behaviors are organized and controlled by perceptions of capabilities—the kinds of behaviors I have at my disposal and whether they can objectively be applied in a given context (Are there sufficient similarities between the situations so that the behavior might naturally generalize to that context?).

The awareness of behaviors and capabilities can be understood in terms of the functions of the frontal cortex and contextual framing by the interaction of pattern matching in the hippocampal formation with conscious outcomes or sensory experience represented in the frontal cortex (DLPFC, OFC, VMPFC, ACC, and Anterior Insular Cortex (AIC)) (Bechara, 2010; Craig, 2009; Chambers et al. 2007; Feil et al, 2010).

Capabilities are arrayed in terms of priorities and preferences according to context. These contexts may represent schemas—what is done in specific contexts, but they also represent arrangements of preferred behaviors, more probable behaviors. This is one of the functions of the midbrain dopamine system, creating hierarchies of preference and salience or importance among behaviors and environmental stimuli.

Capabilities are ordered by the principles of importance/salience reviewed in the section on salience hierarchies. For each capability there is a history of efficacy in various contexts that affects its likelihood of reappearing there. Efficacy comes into play as part of their valuation. Values refer to the level of success a given capability achieves as well as the level to which it becomes available across contexts. On a raw behavioral level, beliefs are generalized subjective reflections of the value, utility and contextual fit between a capability and a context that may include abstract applications of that capability into new and untried contexts. They are also framed by higher-order beliefs about what is appropriate and inappropriate; what can and cannot be done. Capabilities are arrayed and controlled by values, preferences and beliefs.

On a separate level, beliefs about and evaluations of capabilities can be internalized from external sources. This is the essence of Bandura's (1997) social learning theory; we internalize the patterns observed in our models and apply them as if we'd had the experience ourselves. Berger and Luckmann (1967) referred to such internalizations as 'recipe knowledge'. Recipe knowledge is not based on personal experience, but we accept the definitions imposed on us from without. In NLP we have understood such beliefs in terms of acting "as if" (Bandler and Grinder, 1975). This is also one of the important ways in which extrinsic motivations are converted into strong-if-not-genuine outcomes. Bechara (2005) describes these memories, learnings and imaginings as an essential part of the behavioral control mechanism centered in the frontal lobes.

Separate sources of beliefs about capacity flow from perceptions of their consistency with our self-definitions at the next higher level (Identity) and their congruence with personal experience. These are ecological controls on beliefs, values and actions.

Identity flows from multiple sources. It is, however, most firmly rooted in the things that we do consistently, how we value them and the beliefs that we have about them. Although identity beliefs are assembled, on the most basic level, from the self-evident data of experience, powerful transformations of identity can arise from transformative experiences. In such cases the new experience transforms identity. In these cases, the new identity reorganizes the other levels of experience so that they are evaluated and accessed in accordance with the new identity. This is a part of the phenomenon described by William Miller (2004) in his discussion of quantum change. Milton Erickson (1954) called them 'whole life reframes'. They are typical of conversion experiences. For our purposes, a sufficiently powerful restructuring at any of these levels can powerfully affect all of the layers below.

Miller indicates that quantum changers, whether their experience was mystical or more cognitive, came to the conclusion that they were no longer drinkers or druggers. Their experiences at the level of identity–or perhaps at a transcendent spiritual level—transformed their self-definitions which in turn affected the salience of various behaviors. They had not decided to change the behavior; they no longer occupied an identify that incorporated the behavior.

It is important to realize that in Dilts' model, a dramatic change in any level above the problem behavior can cause changes in preference, beliefs and values at all of the lower levels. We have referred to these changes as reframes.

If we consider the Vietnam era GIs who developed addictions while overseas, we may assume that for some, the return to America provided a reassertion of a previous identity that was sufficiently powerful to reframe preferences, beliefs and values so that what happened in Vietnam could stay in Vietnam. As no context was sufficient to re-evoke the identity assumed in Vietnam, heroin use dropped from the behavioral horizon as old identities and preferences reasserted themselves. For others, the shift in context was sufficient to create a change at the level of preferences, beliefs and values. As the context of war was now gone, this meant that the capability—adjustment to life threatening stress—represented by heroin use was neither needed nor valuable. It could be out framed by other behaviors which became more salient in the contexts of home, community and opportunity.

In the study of morphine-using pain patients, the use of the drug appears to have been determined by environment. In this case the environment was pain. When the cues for use ended, so did the use.

These changes represent separate neurological circuits that are either part of the addiction related network or are separate from it.

Insofar as a network of perceptions and behaviors can be evoked that is separate from problem contexts and sufficiently salient to make those choices attractive, the new behavior will come to dominate behavior. Sometimes this occurs with the reemergence of traditional or suspended patterns and at other times the patterns are emergent properties of the individual's life experience (Chambers et al. 2007; Peele, Brodsky et al., 1992).

In a previous section on diagnosis we noted that the first level of addictive spectrum disorders is characterized by problems in specific places or with specific persons. This stimulus bound level of substance abuse can be resolved by such simple measures as doing things differently, avoiding certain places or people or by taking a different path. They are out framed by behaviors; what can be done. Over time, these choices become habitual and the problem behavior fades.

John Walter, one of the originators of Solution Focused Brief Therapy, tells the following story (personal communication, 1995). One of his clients was a working man who had developed a serious gambling problem. He regularly spent his entire paycheck, and had from time to time spent other money he could not afford, at the Off Track Betting parlor. As Walter delved into the structure of the problem, he discovered that the man only gambled at one OTB parlor and that that parlor was just outside of the subway line that the man took to work each day. Walter asked the man what would happen if he took a different route. He claimed that he didn't know but with the therapist's permission, and at his urging, the man began taking a different subway to work. The problem disappeared and never returned. Here is a case where permissions about capabilities allowed a man to make new choices about behaviors that resulted in freeing him from a contextually bound problem.

In terms of the neurological patterns detailed by Chambers et al. (2007), the new path to work, asserted a different set of circuits in which gambling was neither accessible nor conceivable. The new pattern did not include the option to gamble.

At other times, the problem is a behavior that might be out framed by other choices and other capabilities. We have all known people who have become concerned with certain behaviors and made a determined effort to change them. One informant relates that in the 1970s, he was smoking three packs of cigarettes a day. At that time, the price of a pack had just risen to the unthinkable level of fifty cents per pack. At one point, he decided that it was time to quit both to save money and for health reasons. He noted, however, that he needed something more than just a casual decision to stop smoking. Being at that time a committed behaviorist, he decided to associate smoking with physical discomfort and to use that as a starting point for abstinence. One night, he bought several packs of cigarettes, made lots of coffee and stayed up all night smoking cigarettes and drinking coffee until he felt exhausted and quite ill. As the sun rose, he decided that he would have one more smoke and that would be it. He reports that he has never sought another cigarette.

In this case, a decision was made based on the assertion of two positive outcomes that had risen to the level where they could challenge the value of smoking: better health and more available money. This led to a reassessment of his personal capabilities, the behavioral choices available to him about smoking. He decided, based on his valuation of health and money above cigarettes, that he would use other capabilities to change his relationship to smoking. The revaluation of smoking led to the reordering of capabilities in order to stop smoking.

In another example, two members of a family known to the author became aware of the fact that they were drinking more than they ought. Both had noted that alcohol was getting in the way of other behaviors and was beginning to damage the quality of their interpersonal relationships. Based on the fact that they valued a sense of personal control more than alcohol and that their interpersonal relationships were more important than drinking, each made the decision to stop drinking and to stop going to the places were drinking was the central focus. Both successfully overcame their drinking by taking advantage of other capabilities. In both cases, the decision was driven by the existence of values that were more salient than drinking and could be used to empower the alternate capabilities that was the change itself.

Gray (Personal Communication, 2008), tells of a client who had a serious addiction to cocaine. He had been through most of the official programs offered by the federal government and had been through four different probation officers. As a last ditch effort he was assigned to Gray. Gray asked the simple question, "What do you need in your life so that this will no longer be a problem?" After some hesitation (several weeks), the client indicated that as a child he had been an altar boy and had always wanted to be a Catholic priest. He said that if he went back to church, cocaine would cease to be a problem. After several weeks of discussion using NLP-based linguistic challenges to his resistance surrounding church attendance, he finally brought himself to go to church. Within a few weeks' time he became a regular figure at confession and mass and went from regular chronic cocaine use to total abstinence.

Here, a suppressed identity reemerged in a social context and reordered priorities, preferences and values. Because his spiritual identity, once recovered, was much more important to him than his cocaine use, the reordering of preferences and values led to a change in his capabilities and behaviors. The resurgent identity reframed his beliefs and preferences so that he could make alternate choices.

In a similar vein another client had used speedballs—a combination of heroin and cocaine—at the rate of $300 per day for more than six months. One day she discovered that she had become pregnant. Over the course of the next three days she quit completely, experienced little or no withdrawal and did not return to drugs until her daughter was one year old. Here, the identity of Mother powerfully out framed her preference for hard drugs. She said that care for her unborn child was much more important to her as a mother than anything else. When the identity of mother faded, after a year of child care, the identity faded and drugs reasserted themselves.

Chapter Seven

Stages of Change and MET

In 1979, James Prochaska published his epic study of systems of psychotherapy and the kinds of treatment espoused by each. As a consequence of that work, he identified several stages in the process of change. He also identified specific therapeutic techniques that followed from each theory and the techniques that were especially useful during each stage in the process of change. In the ensuing years, he and several colleagues applied the process to multiple areas of health change including diabetes, pap tests, smoking cessation, cocaine addiction and other behaviors with extraordinary success. To date, the perspective has become one of the most well-researched and best-supported approaches to change.

The Trans-Theoretical Model (TTM), or the Stages of Change Model, as it came to be known, holds that persons going through changes move through five definable stages on the way to termination in a stable state that represents their goal. Each of those stages is associated with a set of tasks that must be mastered before moving on to the next stage.

The stages are:

- Precontemplation, where the prospective client is either unaware of the need to change, unwilling to change or uninterested in change.
- Contemplation, where s/he is considering the possibility of change and actively weighing the pros and cons of the problem behavior. In this stage the changer is still open to the problem behavior.
- Preparation, where s/he has made a decision to change within a certain time frame, is planning a strategy for changing and has perhaps already failed in several attempts to change the behavior.
- Action, where s/he has made a commitment to changing and has actually made progress towards the goal of sustained behavioral change;
- Maintenance, where the changer has made sustained and successful efforts at change for a period of at least six months and has undertaken the work of creating and living out an identity that is not oriented to the problem behavior.
- Termination, where the change is complete and the changer no longer identifies with the problem behavior but has assumed a new identity without relation to the problem. (DiClemente, 2003; Prochaska, DiClemente and Norcross, 1994).

The tasks involved in each stage have the effect of moving the 'decisional balance', the level to which the individual is committed to change, away from the problem behavior. This is done through the consideration of change, the planning of change, changing and establishing behaviors that will maintain the change. The approach is incremental. Moreover, it applies to both changing for good—moving away from negative behaviors as well as changing for ill—such as initiating substance abuse. The outlook is significant in that it assumes that change is a *process* and that the *process* takes time. Unlike classical approaches to change, it understands that slips, backslides and relapses are an essential part of the process. As a result, it holds that changers may go through the cycle of changes multiple times before succeeding.

In the model, the movement from precontemplation to contemplation is mediated by consciousness-raising, dramatic relief and environmental reevaluation. Consciousness-raising is information gathering by reading, discussion and self-observation. It is important in that it allows the client to assess whether there is a problem. Dramatic relief may involve role playing and emotional enactments of the realities of the problem. Environmental reevaluation looks at the ecological effects of the problem—how others are affected. All of these initial strategies are educational on both cognitive and emotional levels.

The transition between contemplation and preparation is mediated by Self Reevaluation. Self-Reevaluation is the process of learning how one actually feels about oneself in relation to the problem behavior. It may involve corrective emotional experiences, and reevaluation of goals, meanings and values.

The movement between contemplation and action is mediated by Self-liberation. This consists of making a commitment to act or of framing the belief that *one is capable of changing in a meaningful way*. It can include decision making therapy and commitment enhancing techniques such as letting people know that you've made a decision to change. One of the techniques suggested for this stage is telling people about the decision to change.

Finally, the movement from action to maintenance includes such things as reinforcement management, helping relationships, counter conditioning and stimulus control. Reinforcement management refers to setting up opportunities for reward as one does the right thing. It may include personal rewards, community reinforcement, contingency contracts and other means of reinforcing positive gains. Helping relationships provide opportunities to talk through problems and receive advice from others. They may include self-help groups. Counter conditioning includes structuring alternatives for problem behaviors, desensitization, assertions and positive self-talk. Finally, stimulus control includes avoiding or countering stimuli and situations related to the problem behavior. Restructuring the environment and avoiding high risk situations are typical strategies. It may include submodality manipulation (Gray, 20008; Prochaska, DiClemente & Norcross, 1992).

In this model, precontemplation, although occupying the same cognitive space as denial, is not equated to denial as it is in more classical formulations. Rather than taking the position that a person *should* know that there is a problem and that their denial is as much a *willing* failure to take responsibility as anything else, the model holds that many people are unaware that their behavior is problematic and need to be made aware of it in a gentle and respectful manner. This is also called consciousness-raising. It understands that a person's relationship to change is ambiguous, especially in the early stages. A person who is very interested in changing may still continue dabbling with the problem behavior until they have developed significant experience with other options.

In a further break with classical models, the Stages of Change moves towards a shift in identity. Persons who have overcome addiction spectrum disorders are not seen as being addicts in remission, but are viewed as normal people who may have had addiction spectrum disorders in the past. The end state is seen not as a suspension of the problem behavior, but the creation of a new identity, or perhaps the reassertion of a more fundamental identity with no reference to the problem behavior.

One of the signal insights that emerges from the Stages of Change Model is that all of the change from Precontemplation to Action, all of the behavioral transitions from being unaware or unconcerned about the problem, to actually taking action and doing something about it, is determined by one thing: a shift in the perceived value of changing.

According to Prochaska (2003; DiClemente, 1994; Prochaska et al., 1994), by 1994, the model had been applied to changes that ranged from diabetes glucose monitoring, through weight loss and on to the cessation of tobacco and cocaine use. An examination of thousands of records concerning the application of the model to these diverse categories of change found that in virtually every case, the changers had one thing in common. In each case the breakthrough occurred when the changer identified a positive health outcome— something that they wanted—that was more important and more meaningful to them than the problem behavior. When such an outcome was identified, it empowered movement from precontemplation to action, often collapsing those stages into a very brief interval.

This is the strong principle of change: *wanting* the new health behavior is far more important than *not wanting* the problem behavior. In fact, the capacity to perceive the problem behavior *as* a problem behavior grows in proportion to the intensity of the desire for the new, positive behaviors. People do not change because they see the problem, they change because they find something more valuable than the problem; then, they see the problem (Prochaska, 1994; Prochaska, et al., 1994).

In the context of change more generally, this same insight could be understood as hope for something better, motivation to become something more. In the strong sense used by Prochaska, it represented hope for better, healthier behavior. In a broader context there is the strong possibility that hope generally, especially as an expression of the ideas of personal growth and development espoused by C. G. Jung and Abraham Maslow, might propel persons through change in a way that they would find meaningful and personally motivating. In a more general sense, Prochaska's insight is a strong validation of the material on intrinsic motivation and the crucial role that positive outcomes play in change of any kind (Gray, 2005, 2011b; Maslow, 1970).

The primacy of an anticipated future for this perspective is often overlooked as practitioners work through the stages in a rigid fashion, as presented. Prochaska is emphatic about the importance of the future orientation; that *vision*, is responsible for motivating most of the progress in change. Looking a little deeper, we may also understand that one of the reasons why people cycle through the changes more than once is that they have inadequately specified their outcomes (Prochaska, 1994; Prochaska, et al., 1994).

The next chapter is devoted to a discussion of the well-formedness conditions for goals and outcomes and may be used to maximize the utility of Prochaska's insight.

Arising in parallel with the Stages of Change model is a set of techniques called Motivational Interviewing or Motivation Enhancement Therapy (MET). The central points of Motivational Interviewing were set forth over a period of years by several luminaries in the field of addiction studies including William Miller and Carlo Di Clemente. Although MET is not an NLP-based approach, it is the one scientifically validated approach that we are most likely to encounter and so deserves our attention. What follows is only a brief overview with suggestions of how an NLP perspective could make it a more productive approach (Miller, 1995; Miller, Zweben, DiClemente, & Rychtarik, 1994; Treasure, *2004).*

Motivational Enhancement Therapy (MET) begins with an intake interview and an assessment of how and when the client uses the problem behavior or substance. In addition to a drug history, or a history of the problem behavior, the interviewer elicits from the client, an indication of their highest values and most important outcomes. It is essential to the process that the client identifies some positive goal or outcome which is necessary to empower the anticipated change. During the interview and subsequent sessions, emphasis is placed on the contrast between the client's positive goals and outcomes and the effects of using drugs or alcohol or other problem behaviors on their progress towards those goals. Simultaneously, the client is asked to talk about the places where he or she is already having success and how that relates to the problem behavior. By enhancing the contrast between the successes while abstinent and the interference created by the problem behaviors, the sessions seek to establish ambiguity about the utility of the problem behaviors and to enhance the perceived value or conscious commitment to the outcome.

According to Miller:

> ... the MET approach addresses where the client is currently in the cycle of change, and assists the person to move through the stages toward successful sustained change. For the ME therapist, the contemplation and [preparation] stages are most critical. The objective is to help clients consider seriously two basic issues. The first is how much of a problem their drug use poses for them, and how it is affecting them (both positively and negatively). Tipping the balance of these pros and cons of drug use toward change is essential for movement from contemplation to [preparation]. Secondly, the client in contemplation assesses the possibility and the costs/benefits of changing the drug use. Clients consider whether they will be able to make a change, and how that change will impact their lives (1995, p.3. Modified as noted.)

In accomplishing these goals, MET uses a set of tools and presuppositions that include: expressing empathy using active listening to convey a real sense of understanding about the client's perspective and needs; developing the discrepancy between the problem behaviors and the client's most deeply held values; sidestepping resistance through empathy and; building the client's self-efficacy beliefs by building confidence in their capacity to change (Treasure, 2004).

For the most part, the root concepts in Motivational Interviewing will be familiar to NLP practitioners. They include rapport as verbal matching (empathy), avoiding resistance by reflecting the client's perspectives (feedback frame), and building positive efficacy beliefs about change. Where it differs from standard NLP practice is in: 1. its focus on the problem and its contrast with the stated outcome and 2. its lack of tools for recognizing well-formed outcomes or enhancing their utility in fostering change.

MET is a perfect place for the integration of several of our signature applications into mainstream treatment contexts. For those who are bound to a more traditional clinical role, MET has several specific places where NLP techniques can be used to enhance treatment outcomes. Some of the following applications will be discussed in subsequent chapters.

Empathy: Where MET often speaks generally about empathic and active listening, NLP has developed a repertoire of rapport skills that have been repeatedly shown to enhance positive affect and trust between the client and the treatment provider. Linguistic matching with the matching of breathing and posture provide strong feelings of relatedness between NLP practitioners and their clients (Asbell, 1983; Brockman, 1980; Day, 1985; Ehrmantraut, 1983; Frieden, 1981; Green, 1979; Hammer, 1980; Palubeckas, 1981; Pantin, 1982; Sandhu, 1993; Schmedlen, 1981; Shobin, 1980; Thomason, 1984).

Feedback Frame: NLP also provides specific tools to ensure that rapport is not broken on the level of content, by using the feedback frame. This is the skill of accurately mirroring back to the client their own meaning in way that they will recognize and find to be an accurate reflection of their own statements.

Well-Formed Outcomes: Despite Prochaska's clear understanding that a meaningful future is a crucial element in changing and that the identification of such an outcome can speed progress through the stages of change, many MET practitioners are unaware of

NLP's criteria for creating well-formed, intrinsically motivating outcomes. The possibility of evaluating and establishing motivating and transformative outcomes is one place where NLP can make a significant contribution to the practice. NLP can help the client to identify well-formed outcomes; outcomes that are sufficiently meaningful that they become transformative. By focusing on the positively desired outcome, the discrepancy between felt outcomes and present practices comes into high relief and the client can begin to actively participate in change.

Efficacy Beliefs: These can be modified using several standard NLP techniques including the restructuring of beliefs using submodality mapping and by providing strong experiences of efficacy by teaching clients how to make changes in their own subjectivity, using simple and direct techniques.

Chapter Eight

Outcomes

Before turning to change strategies, we need to take account of motivation to change and what NLP has come to call well-formedness conditions. These are often key to establishing the kind of motivation that Prochaska points to in the stages of change model. Nevertheless, the simple issue of motivation to change must be considered especially with regard to addiction spectrum disorders.

As was noted in the first chapter, motivation is an important part of assessment. Many people with substance use disorders come to treatment for reasons other than consciously and congruently wanting to change. As we said, many are sent by the courts, others do it to please their relatives and still others come because they think that it is the right thing to do. Even those who come really wanting to stop, need to be trained to find an outcome that is positive, something that they want for its own sake, whose consequence can be the termination of the problem behavior. Our discussion of well-formed outcomes and the generation of motivations is crucial to this determination.

We typically treat motivation as though it were something, as if you could have a pound of it and keep in your pocket. This is a fallacy. Motivation is only had with regard to specific outcomes. If you don't know what you want, you will have no motivation.

If someone is unmotivated to change, approaches to treatment can be framed as preparation for treatment as they provide the essential kinds of reframes that will result in problem cessation. The Brooklyn Program was structured so that it would provide unmotivated participants with the very kinds of resources, decision making capacities and future vision to make changing possible. It did this in such a way that each part of the program produced behaviors that were valued by most participants. They provided an intrinsic motivation. Core Transformation can be used to work on some problem that the client is truly motivated to work on. In the process, the deep, transcendent states evoked can set the stage for more far reaching and relevant changes.

So, as we go forward, it is important to understand that motivations can be evaluated and shaped by the methods we will discuss. It is also important to realize that profound positive experiences can be shaped into motivations for change which may not have existed before.

The idea of well-formedness conditions for goals or outcomes is a central element of Neuro-Linguistic Programming interventions. It developed more or less directly from the work of Noam Chomsky. Just as Chomsky held that native speakers of any language can intuitively identify whether a communication is well-formed or meaningful, so, human behaviors require certain kinds of structures in order to make them meaningful, motivating or efficacious. Typically these conditions include the specification of the formal characteristics of the elements and their required order (Bandler & Grinder, 1975; Dilts, & Delozier, 1990; O'Connor & Seymour, 1990; Linden & Perutz, 1998; Dilts & Delozier, 2000; Gray, 2008, 2011b).

At their most basic level, the NLP well formedness conditions for any given outcome specify that:

- The outcome must be stated as a positive thing or experience; something wanted, not something unwanted or ending.
- The outcome must be something that is under the goal seeker's personal control which also implies that the task should not be stated too broadly.
- The outcome must be specified in terms of sensory experience; it must be described in terms of what can be seen, heard, felt, tasted or smelled.
- The outcome should be evaluated for ecology: what it will change in the person's life and the lives around them.
- The outcome should be imagined and experienced in fantasy as fully as possible (Andreas & Andreas, 1989; Bodenhamer & Hall, 1988; Cade & O'Hanlon, 1993; Dilts & Delozier, 2000; Miller & Berg, 1996; Linden & Perutz, 1998).

It is noted that, for the most part, these characteristics are typical of deep, intrinsic motives. Intrinsic motivators are desired positively (Deci and Ryan, 2008). They are characterized by choice and personal autonomy; they often include strong self-efficacy beliefs

(Baumeister and Heatherton, 1996; Deci & Ryan 2008; Hulleman et al., 2008; Koestner, 2008; Nootz, 1975). Because they are often rooted in previous or vicarious experiences (Baumeister and Heatherton, 1996), they can be specified in sensory terms (often with special emphasis on kinesthetic elements—this is how I will feel). The imposition of well-formedness conditions can often be used to differentiate between extrinsic outcomes with relatively superficial motivations and intrinsic motivations which provide stronger sensory and motivational cues.

During 1992, the author was teaching psychology at a local Community College. As part of a lesson on motivation, he asked students to apply NLP well-formedness criteria to outcomes that they had already set for themselves. An important facet of the exercise was the imaginal experience of the anticipated outcome. That is, after specifying a positive outcome, after determining that the outcome was under their personal control and specifying several means by which the student would know that they had attained the desired state or position, they were asked to imagine stepping into the end state and trying it on.

On this occasion there was a young woman in the class who had been working towards a degree in nursing. She had just begun the program and had no idea of what it was that a nurse actually did. When she tried on the imagined experience of the day-to-day realities of nursing--the suffering, the filth, the blood--she came rather quickly to the realization that it was not something that she wanted to do. She changed her major soon thereafter.

When we begin to consider drug treatment in terms of well-formed outcomes, we immediately run into a serious problem. Most treatment strategies are centered on an ill-formed outcome—stopping the unwanted behavior. Not using drugs or alcohol anymore is not and cannot be a well formed outcome. Let's examine this.

A well-formed outcome must be stated in the *positive*, in terms of a positive goal; **what I want, not what I don't want**.

In 1987 Daniel Wegner and his colleagues published a study entitled Paradoxical Effects of Thought Suppression (Wegner, Schneider, Carter & White, 1987). In that now famous study, they asked students to begin by reporting into a tape recorder everything that came to mind during a five minute period. After five minutes they were given further instructions. During the next five minutes they were instructed ***not*** to think of a white bear. If they did, they were to name it on the tape recording and ring a bell. After this period, they were told to just speak into the microphone making no special effort. If, however, they thought of a white bear, they were to report it and ring the bell. Needless to say, the effort to *not* think about white bears primed the participants to do precisely that—think of white bears. Moreover, in the third period when they were no longer instructed to not think of a white bear, they also thought of them more often.

The authors comment that their data

> … indicates that the task of stopping thoughts has the effect of producing associations of that thought with many other thoughts immediately available to the person, and that these associations function to make the thought rebound when the injunction to avoid the thought is no longer in effect. (P.9)

The relevance here is that negative outcomes often have paradoxical effects. Many authors have pointed out that the brain does not compute negation. Whenever a negative proposition is suggested, the problem must first be represented and then erased, faded-out or otherwise changed. A negative outcome always ends up enhancing the perceived importance of the problem object or behavior. In the case of addiction, attempts to suppress the urge or not think about the problem only tends to increase its potency.

Beyond these observations, positive goals have qualities that can be imagined. They can be seen, moved towards, and manipulated. They provide a focus for attention. Negative goals are much more diffuse. As their focus is negative, they can lead anywhere so long as it is away from the object (Gray, 2008).

As noted earlier, Kringelbach (2007), in his review of the function of the OFC, reports that stimuli are perceived as more motivating in proportion to the depth of their sensory representation. This suggests that the more fully represented an outcome is, the more compelling it becomes, a point made repeatedly by the literature of NLP. Not doing something, by definition, cannot create a strong image (Bandler & Grinder, 1979, Dilts, et al., 1980, 2000; Linden & Perutz, 1998)

Our problem with addictions treatment outcomes continues because most people with drugging and drinking problems don't know what they would *rather* have. More often than not, when pressed for a positive alternative to *not using* problem substances, they will provide a stock answer: money, power, fame or sex. Just as frequently, they provide the answers that they think they should. For others, when asked to create a positive outcome, the answer tends to be relatively content free—I want to live a healthy life style; I want to be healthy; I want my family back; I want to live my life. When asked what that means, they often reply with the negative definition—I won't be doing this anymore. Sensory specificity, concreteness, is a crucial part of realizing any goal. If there is no way to test that you've attained it, you do not have a well-formed outcome.

A third problem with the outcomes inspired by standard treatments is that they are not under the client's control. If stopping the problem behavior were a simple matter of will power, just stopping, they probably would have done so long ago. As has already been noted, the salient feature of addiction spectrum disorders is precisely loss of control. As a result, the outcome—no longer doing this—is ill formed on another level; for the most part, it is not under the control of the client.

This dimension is doubly complicated because of the natural impulse to assume that the problem behavior is completely under the addict's control. This is the fundamental attribution error, the belief that if I have a problem I have a reason for it. If, however, you have a problem, there is something wrong with you: you must be bad, broken, or crazy. All of us have known people who just stopped one behavior or another. Many of us have had the experience of just stopping. What makes the difference here is that all of those changers probably had a positive outcome that motivated them to stop because the problem behavior interfered with the more highly desired outcome (Gilbert & Malone, 1995).

You may recall that the Stages of Change Model expects that the client will relapse. Part of the reason for the relapses may be that most clients have developed no strategy for creating well-formed outcomes. Any outcome—especially *not doing this anymore*—is believed to be acceptable. As a result, the client must cycle through multiple trials before they find an outcome that is well–formed and sufficiently motivating to impel them through the stages of change.

There are other problems with the motivational structure of classical addictions treatment. The outcomes tend to be content free. This may be because, for most people caught up in such problems, there is often no positively desired outcome, and, where there is, it tends to be nominalized—empty of sensory data. In such circumstances it is nearly impossible to set a sensory specific goal other than in terms of the non-performance of the problem behavior.

Ideally, an outcome for ceasing addiction spectrum disorders should have nothing to do with the problem itself, but should be intrinsically motivating, highly pleasurable and should work to strengthen all kinds of behaviors that mitigate the problem behavior. These might include a specific personal attainment (getting a degree, fulfilling a dream, and living up to a spiritual aspiration).

There is another complicating factor in classical addictions treatment. When outcomes are stated they are very often extrinsic and reflect the relatively impoverished backgrounds of the people in treatment. Extrinsic motivators include things like money, power, sex and popularity. Because they have lived with little hope, few have created visions of a meaningful future.

Levels of Motivation

While well-formed outcomes can be crucial elements of any motivational strategy, it is important to understand that there are multiple levels of motivations and motivators and not all of them will produce transformative change. Briefly, there are large motivations whose contexts are life-long. They include identities and callings. There are motivations that are driven by smaller contexts. These are situationally determined goals and outcomes at work, at school and at home. There are motivations that are relatively evanescent; they arise and dissipate in short order. Finally, there are motivations that disappear unless they have significant support from others.

The longest lived motivations are typically about growth into identity. Maslow spoke of these life-long motivations in terms of the actualization of personal potential and developed a psychology of Self-actualization. Jung called it Individuation and spoke of realizing the unconscious capacities of the Deep Self at the level of full consciousness. In most cases, persons who are self-actualizing or individuating have learned to bring all of their activities into the service of the life calling. For these people, most motivations serve, or have been subsumed under, the larger purpose of the life calling or path of Self-actualization (Gray, 1996, 2008; Hillman, 1996, Maslow, 1970).

These motivations may develop early or late. They may arise out of spiritual experiences or Quantum change. They may arise out of an increasing number of coincidences, synchronicities or peak experiences that lead an individual inexorably in a certain direction. In general, they develop a pattern of flow experiences that confirm the goodness of fit between the individual and the life that surrounds her. A life ordered by this level of motivation answers to M. Scott Peck's definition of spirituality as "knowing one's place in the universe" (Csikszentmihalyi & Csikszentmihalyi, 1988; Gray, 1996, 2008; Hillman, 1996, Maslow, 1970; Miller, 2004; Peck, 1998).

One of the important features of these motivators is that they often meet the conditions for well-formed outcomes without the external application of the criteria. They also foster experiences of flow as a natural part of life experience.

For people who are not developing towards conscious realization of a calling, the longest lived motivations serve various roles or identities that are tied to contexts such as employment or family. When the role ends or nears its end, we find people who are devoid of direction. This is a partial explanation of mid-life crisis, empty nest syndrome and why it is that many people who retire often become ill and die soon thereafter. The roles that have given them identity and meaning have dissipated and, having no root in themselves, they are forced to find new meanings or die (Hillman, 1996).

These motivations typically involve internalizations of external values and definitions that may be only superficially related to the individual's full potential or life calling. They can be structured by well-formed outcomes, but do not generate well formed-ness conditions of their own accord. That is to say that the person may accept the life path, and even if it provides a great deal of reward, it is always felt to have 'missed the mark'.

Some motivations are centered about attempting to recreate, relive or maintain a past identity. The *Al Bundy Syndrome*[5] represents the motivations of the person whose glory days were in High School, College or in the Armed Forces and all of their life is spent grasping at a past that can never return. While more grounded in pathology than the more practical motivation of the working man, it likewise ends in disillusionment and death.

Some motivations are based on imaginal manipulations of meaning. Advertisements make use of sensory submodalities and conditioning to awaken motivations to buy, get and have certain

---

[5] This was named by my wife and editor, Florence Tomasulo Gray, and I after noting how well the problem was echoed in the 80s television series, "Married with Children."

things. By mimicking the archetypal numen of spiritual, transcendent and life-critical reinforcers and stimuli, they awaken the immediate desire for things we may not really need or actually want (Andreas & Andreas, 1987; Bandler & LaValle, 1996; Bandler & MacDonald, 1987; Dilts, et al., 1980; Dilts & DeLozier, 2000; Goodwyn, 2012; Haule, 2011; Gray, 2008, 2012).

It is not uncommon for some unscrupulous NLP practitioners, often involved with sales, to manipulate the submodality structure of an object that they want to sell, or an objective that they have for someone. By subjectively making the object bigger, brighter, closer; turning on the sound and turning up the volume and focusing on the locus of the feeling generated in the subjective experience of the object, it can be made highly desirable. Such motivations, unless they are congruent with deeper values, often fade and end with the realization by the victim that they have been manipulated or deceived (Andreas & Andreas, 1987; Bandler & LaValle, 1996; Bandler & MacDonald, 1987; Dilts & DeLozier, 2000).

In Early Modern English such manipulations were known a 'glamours' and were often part of a magical repertoire called *grammarie*. According to the *Oxford English Dictionary*, this word originally meant a spell, especially a spell that affected vision. It was often associated with Fairy Gold which, after being in human possession until dawn, was often found to have turned into coal, ashes or dung.

ASSESSING AND CREATING WELL-FORMED OUTCOMES

In order to create intrinsic, well-formed motivators that are sufficiently powerful to support transformative change, motives that are capable of awakening the kind of life changes observed by Miller in Quantum Change and by Erickson in his whole life reframes, and even to speed clients through change work, it is important to be able to assess those motivations and provide alternatives where necessary.

There are several means for identifying intrinsically motivating outcomes or structuring intrinsic motivation. The same techniques can also be used to move a client towards the identification of more intrinsic motivations when they have settled for more superficial options. In each case, the NLP well-formedness conditions for outcomes may be applied to provide greater specificity and a real-time evaluation of the utility of the outcome itself. Beyond the application of the well-formedness conditions, here are three techniques for assessing, enhancing and creating intrinsic well-formed outcomes.

- **Asking**; what do you really need or how do you need to feel in order for this to work for you?
- **Seeking end state energy** in the present. Experience now, in the present how you anticipate that you will feel when you have this outcome.
- **Finding primary outcome sequiturs** or Core States. If you have this, what will it do for you on a much deeper level that is intrinsically more meaningful and more valuable for you? If you have that feeling, what kinds of outcomes will you be seeking? Are they the same or are they different from those you imagine now?

Tools for creating intrinsically meaningful outcomes

Asking

One of the important ways of creating positive futurity is by the identification of a future outcome that is already meaningful to the client and connects to a deep sense of personal identity. From time to time a client will come to you and will be able to articulate the outcome immediately. This is based on the idea that many people know precisely what they need in order to change, they just don't follow through (Bandler & Grinder, 1975, 1979; Bandler, 1993, 1999).

In the late 1990s, the author had a client who knew precisely what he needed to do in order to stop using cocaine. He was an ex-Mafia lawyer who had been through multiple treatments and multiple probation officers. He was assigned to the author as a last resort. In the initial interview the client was asked whether he knew on any level what he had to do in order to stay clean. What did he need to do or become in order to overcome his problem?

In response, he became quite emotional and said that he knew but didn't want to say. Over the next several weeks the question was pressed. Eventually he replied that he needed to get back to church. He had been an altar boy as a child and had considered becoming a priest. If he could return to the church, he said, his problems would be over. Over the next several weeks we wrestled with the Meta-Model questions, "What prevents you from doing that?" and "What would happen if you did?" "How will you feel when you've done that?" After a few more weeks, he took the plunge; attended confession and began regular church attendance. Very quickly his cocaine use slowed significantly and, after a few weeks, stopped altogether. To the author's knowledge he remains drug free to this day.

In this case there were two important criteria. 1) The outcome was deeply valued by the client—an intrinsic outcome, and 2) He recognized it as a defining element of sober (pre-drug or positive future) identity. In terms of Jungian theory we can understand this as a manifestation of the Deep Self, the centering and directing archetype. The Self-archetype draws the individual towards wholeness and is often expressed in the best forms of the spiritual and religious impulse. In some sense its draw can be understood as heeding a call (Hillman, 1996). By awakening it as a possibility and fostering its realization, the client recreates a condition of life that is strong enough and stable enough to compete with the addictive urge. One of the important distinguishing characteristics of intrinsic outcomes is that they are not about possessions or status; but about identities and actions (Gray, 2011b, Deci and Ryan 2008).

On a neurological level we can return to the observations of Chambers et al. (2007) to the effect that the behavioral schemas that hold the key to sobriety are often completely separate from the substance abusing life style. In this specific case, they were separated in the classical sense in which a religious, holy or sacred space is literally 'set apart.' When the client stepped into that space, there was no more place for substance abuse (Gray, 1996; Haule, 2011).

For this kind of inquiry, NLP has developed specific linguistic tools in the Meta-Model. The Meta-Model consists of a systematic listing of the distortions and deletions that characterize everyday language. They represent the transformations of experience that simultaneously make day-to-day living possible, and that prevent us from seeing much of what we need to see in order to grow and evolve

(Bandler & Grinder, 1975; Bandler, 1993; Bodenhamer & Hall, 1998; Dilts, Delozier & Delozier, 2000; Lewis, & Pucelik, 1990; Linden & Perutz, 1998; O'Connor & Seymour, 1990).

The Meta-Model was developed from linguistic patterns rooted in John Grinder's work with Chomsky's transformational grammar. In general, it is presented as a list of patterns about language that indicate that the speaker has deleted, distorted or generalized his representation (the surface structure) so that some important datum of experience that would otherwise appear in a full representation (the deep structure), is now missing. The application of Meta-Model Challenges is designed to open limits and reveal presuppositions and can be very useful. There are several excellent texts on the meta model including The Structure of Magic (Bandler& Grinder, 1975) and The Magic of NLP Demystified (Lewis & Pucelik, 1990)

End state energy

John Overdurf (2008) describes the utility of what he calls end state energy. He begins by pointing out his observation that the most crippling linguistic pattern in all of human experience is this: "If only I had x, then I would feel y." Most practitioners of Neuro-Linguistic Programming will recognize this as a cause-effect violation; the mistaken idea that people or things cause our feelings. Using this construction, we miss the point that we are the source of our own feelings.

The main problem with this kind of thinking is that it cedes control over our emotions and feelings to someone or something else; something outside of us and beyond our own control. One of the more important problems with this style of thinking is that it is often a characteristic of extrinsic motivators. Like the young lady who had an abstract image of a nurse as a respected care-giver: her motivations for the specific outcome were extrinsic. The one thing that she knew she wanted was the feeling associated with the position or role.

Whenever people engage in this kind of planning they tend to create a self-defeating loop. Overdurf calls it deductive motivation. We begin with a low energy state and hope to multiply that energy as we struggle to maintain a single minded drive to reach that outcome. Because, however, we have imagined that all of the positive value lies in the future, we are defeated by a kind of psychological law of entropy—our energy is exhausted before we get there. Recall our earlier discussion of Baumeister and Heatherton's (1996) idea of self-regulation failure. If all of our energy is devoted to an abstract outcome, we may not be able to get there from here.

There is, however, another possibility. Overdurf calls this inductive motivation. Here, we begin with a present time experience of the desired state—the end state energy—and apply it to the 'smallest next step' in the path to our goal. Because this energy is already ours, because it is already attached to our imagined outcome and because it is aimed at a small step that will end successfully, it tends to become self-sustaining.

Overdurf points out that in order to imagine that future feeling, we must already know what it is and how it feels. His solution, much like Connirae Andreas' strategy in *Core Transformations* (1994), is to ask the client to fully experience how they would feel if they had that experience in the present. If you can imagine it, you can experience it.

Here is how the great neuroscientist Antonio Damasio describes it with regard to memories; the same applies to vividly imagined events:

The records we build of the objects and events that we once perceived include the motor adjustments made to obtain the perception in the first place and also include the emotional reactions we had then. They are all co-registered in memory, albeit in separate systems. Consequently, even when we "merely" think about an object, we tend to reconstruct memories not just of a shape or color but also of the perceptual engagement the object required and of the accompanying emotional reactions, regardless of how slight. (Damasio, 1999, p. 148)

On a practical level, this means that if you can think about a state or imagine it, you can reconstruct it and experience it in the present. Insofar as you (or your client) have an idea of how you will feel in that desired future, you can invite them to take a minute and step all of the way into that feeling, **now**. Ask the client to imagine that they can step all the way into the feeling. Have them notice their posture, their breathing, and the subtle motions of their body. Let them notice how the feeling affects the muscles in their face and locate the warmest part of the feeling in their body. Have them notice where it feels the best and instruct them to turn their attention into that place.

This is end state energy. It is already associated with the goal or outcome and is already present and available to consciousness. Holding this feeling in mind and body you—or your client—can now turn your attention to the 'smallest next step' that can be taken to accomplish this outcome. As they do so, have them notice whether that 'smallest next step' receives and multiplies the energy or not. If it does, tell them to "Allow yourself to become more excited and more motivated to continue on the path to your goal. Feel the strength of it, now."

This is, of course, a practical application of the neurology described by Antonio Damasio, Antoine Bechara and others: memory constrains attention and attention in turn evokes the feelings appropriate to the remembered experience. It is also a clear and practical application of the NLP idea of a resource (Andreas & Andreas, 1987; Andreas & Andreas, 1989; Bandler & Grinder, 1975, 1979; Bechara, 2005; Cade & O'Hanlon, 1993; Damasio, 1999; Dilts, et al., 1980; Erickson, 1954; Gray, 1997, 2011a; 2012; Linden & Perutz, 1998).

If you find that the energy is not there, if the end state energy does not match the 'smallest next step' that your client has imagined, let them step all the way back into that energy, enhance it by directing them to notice where it is centered and how it moves through their body. Tell them to "Spend some time allowing that feeling to spread through you. As you enjoy that feeling, find a 'smallest next step' that will support this feeling. Allow the feeling itself to lead you to the action that resonates most fully with it."

When they have fully associated into the energy, let them use the following three well-formedness criteria to create it as an outcome--while they keep their attention focused on how good this state feels in the present time. Instruct them as follows:

- Pay attention to the feeling so that the feeling guides you to a step or outcome that supports the feeling.
- Make sure that the outcome is under your control and be sure that it is just small enough.
- Make sure that you can specify how you will know that you will have it when you have it.

This technique actually has two effects. On one hand, insofar as the client is attracted to a positive outcome that flows from their own self definitions; something that they desire for its own sake, this technique provides a felt connection in real time to the possibility of getting there. All of the hoped-for affect can be integrated into the implementation or goal acquisition plan as this immediate felt experience is associated into the 'smallest next step' (Baumeister & Heatherton, 1996; Koestner, 2008; Overdurf, 2008).

On the other hand, if the client has started with an extrinsic or relatively meaningless outcome, the process of end state energy can serve to reorient the client towards what is truly important to them. This is usually represented by the felt state associated with the outcome. In this case, the client will step into the state but will find it incongruent with the step that leads to the imagined outcome. When this happens, urge the client to formulate a 'smallest next step' that matches with the *feeling* of the end state which is the more important and truly intrinsic outcome. Engaging in the new direction will be experienced as self-reinforcing and will encourage the client to seek other ends.

Outcome sequiturs

Outcome sequiturs are the real reasons why we do things. They answer the question, what did that really *get* you? They are often the felt consequence of an outcome or, in Overdurf's language, the end state itself (Dilts, Delozier, Bandler & Grinder, 1980; Dilts & Delozier, 2000; Overdurf, 2008).

As noted above, even intrinsic end states, as feelings, are often dissociated from the imagined or stated outcomes owned by a client. Following a chain of outcome sequiturs can often elicit a powerful affective state that can then serve to guide future behavior and to aid in creating appropriately intrinsic motivators.

A particularly eloquent use of outcome sequiturs is provided in the work of Connirae and Tamara Andreas in their work, *Core Transformations* (1996). The process outlined there begins with the elicitation of an outcome. The outcome, whether good or bad, is then made the subject of the following question: If you had that fully and completely, what would you have now that is deeper, more meaningful and more satisfying than just that? How would that make you feel? The client is then given a moment to access the relevant feeling and to make a response. When the new felt sense or objective outcome has been identified and accessed, the question is repeated. For each conceivable sequitur the question is repeated until the client can find no deeper feeling. At this point, the felt sense typically comes to rest in one of several oceanic feelings of spiritual well-being; states the authors refer to as *Core States*.

These Core States can serve two purposes. On one level, if the process begins with the exploration of an urge or impulse, the Core State can become linked to the stimulus that evokes the urge so that the urge now redirects the client to the Core State and the kinds of behaviors that will maintain it. The second purpose, more germane to this topic, uses the technique to restructure conscious outcomes so that they are congruent with the deeper strata of consciousness. In so doing, they create the groundwork for transcendence, that ability to bypass temptations and distractions that draw seekers away from their path. Further, because they represent such a deep level of experience they have the capacity to marshal other outcomes and behaviors in their service. This becomes an explicit application of the systems principles of wholeness and emergence (Gray, 1996; Piaget, 1970). As a result, there is less of a tendency to run short on self-regulatory energy and the behavior becomes self-maintaining (Baumeister & Heatherton, 1996).

On a simpler level, the same basic technique can be used to channel the client's awareness and behavior into more fundamentally valuable intrinsic behaviors. After eliciting a series of outcome sequiturs that end, presumably, at the level of a Core State, the client can be asked whether they have ever experienced anything similar before. An exploration of such similar experiences reinforces both the validity of the present experience and its future accessibility. The client might also be asked to consider what kinds of things would be likely to make these kinds of feelings more available in the future. From here, new outcomes may be structured, rooted in these kinds of states and organized using the well-formedness criteria.

That feelings might be used to structure behavior is implicit in the theories of Peter Lang, who holds that emotional memories and emotional states are represented in the brain as networks of perceptions, semantic associations and actions that create the meaning of the event. According to this model, every emotion implies specific types of behavior with some, like fear, being more fully determined than others. Nevertheless, the more strongly an affect is experienced, the more likely it is to evoke behavioral expression (Lang, 1982, 1994).

Chapter Nine

NLP techniques for motivated clients

You may remember that in Chapter One, we discussed several categories of clients presenting with addiction spectrum disorders. Independent of the diagnostic category that they may occupy, one of the central issues surrounding treatment is whether or not they are self-motivated. As we have noted in much of the last two chapters, intrinsic motivation is an important factor, especially in drug and alcohol treatment.

It has been my position and the position of others in the field that—for the most part—once a person is motivated to change, almost anything will work. Nevertheless, persons who are motivated often need help from outside. The amount of help needed may range from basic permissions to whole-life reframes, but in the case of motivated clients, the assistance may be rather straightforward and may make use of the clients' conscious and unconscious cooperation.

The interventions suggested over the next several chapters are designed only as a representative set and are intended to stimulate thoughts about change. All of these techniques have worked for the author or for reliable informants and all of them are subject to modification in order to meet the needs of your clients. None of them work for everyone; there are no panaceas.

The following techniques will be covered in the paragraphs below.

- Asking
- Permissions
- Creating anchored resource states

ASKING

We have already covered the simple technique of asking. It is suggested by Bandler and Grinder in several of their books and it is still surprising how often it is missed. Treatments of various kinds often proceed based on the written record rather than meaningful communications with the client.

Gray (2008, Personal Communication) recounts the case when a federal offender with a long history of cocaine addiction, complicated by violent psychotic outbursts was referred to him. According to the record, she had attacked prison guards, and was famous for smearing her prison cell with feces. She was allegedly illiterate and incorrigible.

All of this data was presented orally. Before reading the file, Gray invited her in and told her that he had heard some wild things. He indicated that it was important to him to hear what she had to say. The client began to talk and provided reasonable explanations for her unreasonable behavior. Many of her responses, she explained were purposeful actions because no-one would talk to her; everyone relied on the case file. After the initial interview, the offender was seen weekly for long talks about her personal needs and directions. Room was provided for her to discuss her feelings and options for meeting her needs were discussed. She quickly stabilized, remained abstinent–after some initial positive urine tests–and completed her parole without incident.

Although hailed as a near miracle by others in the department, Gray explained that all he did was ask the client what she needed to get things to work for her. She answered, he listened, and things went well. The intractable cocaine use was rooted in frustration. When the frustration was ameliorated, the cocaine use disappeared.

Asking can work, but it does not work for everyone. Crucially, the person for whom asking works typically has outcomes and meaning structures already in place. The answers provided must be answers to real needs with high priorities. Here, the client had a well-formed outcome: she wanted to get her commercial driver's license, a job as a truck driver and set up house with her lover in another state. In order to obtain those outcomes she needed to know the rules and be provided with clear guidance for getting through the system. Once the right questions had been asked and answered, she could get on with her program.

A similar problem was encountered with the Catholic cocaine addict described in earlier chapters. The key to his adjustment was an identity level need, rooted in certain behaviors—getting back to church so he could be a good man—that was incompatible with the successful but ultimately superficial life that he had built for himself. When he finally acknowledged to himself and others that his desire to return to church was real and legitimate, he could allow the more primary identity to re-emerge.

For some persons who have answers, those answers will need to be restructured using the well-formedness conditions for outcomes. They may know broadly what they need but with insufficient specificity to drive action. Outcomes that are not specified in at least three sensory modalities are often insufficient to drive behavior (Dilts, 1983). Where the main intervention is asking and applying meta-model distinctions to help empower change, the application of the well-formedness strategy (see below) can positively impact the change.

Permissions

Like asking, permissions are often overlooked by treatment providers but can be a very powerful means of empowering change. Permission was the crucial parameter in the story of the OTB gambler told by John Walter. As noted, one of his clients had developed a serious gambling problem. He regularly spent his lunch money, his entire paycheck, and sometimes the rent, at the Off Track Betting parlor. Walter discovered that the man only gambled at one OTB parlor and that that parlor was just outside of the subway line that the man took to work each day. Walter suggested that the man take a different train or a different means of transportation to work. At Walter's urging, the man began taking a different subway to work. The problem disappeared and never returned.

One of the root presuppositions of NLP is that if what you're doing doesn't work, do something different. We often fail to understand that we mistake habit and custom for law. We do things the way we have always done because they are habits but justify them as if they were principle-based. At other times we continue in habitual patterns until a manna figure, a person imbued with special power or authority, comes along and gives us permission to do something else.

We have already noted that my father's smoking continued unabated until he received the permission/command from his physician to stop. Sometimes change is that simple: as conditions or permissions change, some behaviors disappear with them.

In the 1990s I had a client who had significant problems with the use of cocaine and marijuana. Fridays and Saturdays were boy's night out and he felt compelled to go despite the fact that he had been arrested several times and his wife was ready to leave him. He was a regular church attendee and never used drugs apart from his friends. His wife complained bitterly about his friends and his drug abuse, but it did no good. When challenged about the problem, he had a string of answers related to manhood and his need to relax. He was concerned about how he would appear to his friends. During our conversations an effort was made to change the frame of his relationships in terms of sex with his wife (was he getting any?), his capacity to 'step up and be a man' and whether or not he had outgrown this particular group of friends. After a period of discussion surrounding these issues, he was able to grant himself permission to separate from this younger group, and take up his role as a provider. In a sense, this permission solved the problem.

In most such cases, meta-model violations—Modal operators of necessity (I must, I have to, I should), Cause-effect violations (If I don't do X, Y will follow; If I do X, Y will follow) and Mind reading (I know they will feel this way)—have the effect of trapping people in behavioral patterns that are less than useful. When challenged appropriately (what would happen if you did or didn't do this? How do you know that they will respond that way; have you ever tested it? Who says they think that? Have you asked?), they often dissolve.

On another level, permissions form a counter belief for the taken-for-granted definitions that arise out of habit. Our beliefs tend to be formulations of what we perceive ourselves doing. I think I am a good man because I do things that may be defined as good. I know I am an addict because I keep using these drugs. Sometimes, permissions break through the glamour of these self-referential loops and allow us to see alternate choices.

Like the answers arrived at by asking, permissions assume that there exists sufficient structure in the person's life so that the new behavior may be appropriately contextualized. When they work, they often work because of a preexistent meaning structure in which the new behavior can find an ecological space. Permission to stop smoking is useful and meaningful, if the smoker has a positive reason for making that choice. If he has a growing career, a healthy family and a desire to live beyond 50, the permission may be sufficient. If the gambler has reasons to live that can make gambling irrelevant, and the habit is grossly contextualized, a permission to do something else may be sufficient to inspire the change.

It should be noted that doing something different after receiving permission to change, might as often as not, be totally unrelated to the problem behavior, even though it may result in a massive impact on that behavior. If the permission makes something that is more valuable than the problem behavior more immediate, more accessible and more intuitive, the increased access to the positive outcome will out frame the problem so that it is no longer a problem.

Once again, the pattern of permissions recalls the small world networks cited by Chambers et al. (2007). Assemblages of perceptions and behaviors are typically associated through preferred nodes in a neural network. When a separate node, associated with a different set of behaviors becomes dominant in organizing behavior (the emergent property), it can give rise to new behavioral patterns. As these alternate behavioral networks come on line, their inherent perspectives make the old behavior inconceivable. This pattern unites permissions with the changes in the Vietnam Veterans, the rat park rats and the patients who used opiates for pain relief without addiction.

CREATING ANCHORED RESOURCE STATES

One of the things that we have noted in earlier discussions is the value of self-efficacy in the context of treating addiction spectrum disorders. We have noted that addictions are by definition problems related to impulse control. In our discussion on motivations, we found that outcomes that increase autonomy are more highly valued than those that do not. In the discussion on well-formed outcomes, we learned that effective outcomes must be under the actor's control. In our discussion of hierarchies we found that things that work are accorded higher levels of salience.

Beyond this, when we look to the information about how outcomes impact problem behaviors, we find that if the new outcome can compete with the problem behavior, the problem may be reframed into insignificance by the enhanced salience of the preferred behavior.

On the level of neurology, we previously discussed the positions of Bechara (2005), Feil and colleagues (2010) and Baler and Volkow (2007) to the effect that there are two facets to addiction spectrum disorders, the inordinate incentive salience (wanting, craving) accorded addictive substances and behaviors and the failure of frontal inhibitory mechanisms either as precursors to or consequences of substance abuse. The following two techniques speak directly to these issues by creating powerfully self-reinforcing behaviors that focus the attention and teach the art of mindfulness. Importantly, they provide a resource that has the capacity to out frame negative behavioral patterns as 'absurd good news'.

The relevance of these techniques, especially the submodality analysis, to addictions is reflected in a study by Bickel, Yi et al. (2011) showing that stimulant addicts typically discounted otherwise highly valuable future rewards in favor of lesser, more immediate rewards. These researchers found that memory training allowed the addicted subjects to respond more normally to the objective value of the rewards. This suggests that mental practice, learning how to focus attention, provides powerful results in changing how people interact with the world around them.

That these behaviors can compete with addictive substances is reflected in a story from Gray (personal communication, 2012). In one of his sessions during the Brooklyn Program a group of skeptical offenders had completed the exercises in a somewhat pro forma manner. When, however, they did the anchoring exercise for the first time, several literally jumped out of their seats and proclaimed, 'This is better than drugs.' More recently (2014) a recovered user of heroin and other drugs discovered the anchoring technique described here and proclaimed, "Now I know what the crack-heads are looking for."

Submodality analysis and manipulation is based on the idea that all subjective behavior can be analyzed into sequences of sensory experience; what we see, hear, feel, smell and taste. On a more finely-grained level, the valence, intensity and meaning of these experiences are determined by the qualities of the sensory experiences. These qualities include their subjective position, distance, intensity, amplitude, stability, focus, etc. The manipulation of these variables can intensify, weaken or change the meaning of an experience. For example, the image of a pleasant memory may be experienced as associated or dissociated, near or far, bright or dim, moving or still, two-dimensional or multi-dimensional, colored or mono-chromatic, etc. Similar dimensions of perception apply to the other senses. Each of these manipulations can change the subjective experience significantly and can, in combination, powerfully impact that experience (Bandler & McDonald, 1987; Andreas & Andreas, 1988, 1989; Gray, 2008, 2011b, 2012).

The second tool, anchoring, is a basic classical conditioning technique that is used to make the experiences developed by shifting the structure of the submodalities transportable and manipulable by the client. In general, it consists of associating a predetermined—though nevertheless arbitrary—gesture as a Conditioned Stimulus with a practiced ecstatic state (developed using submodality manipulation) as a Conditioned Response.

In NLP, anchoring can refer to almost anything from a gentle touch used as a conscious reminder, to a classically conditioned stimulus that evokes a specific, involuntary, emotional or visceral response. Here, anchoring is treated as a classically conditioned learning experience in which repeated pairings of a meaningless gesture with an emotional experience allow that gesture to elicit, and modify, the original emotional experience. These conditioned stimuli may be thought of as triggers for the desired response. They are automatic and relatively immediate (Gray, 2011a, 2011b, 2012).

I recently consulted with a therapist from the North East. She was working with an alcoholic woman who found it necessary to drink heavily for most of the day. Although she had been able to take care of her family responsibilities, by the end of the day she was quite drunk and was finding that her ability to function was decreasing daily. The therapist called me and I provided her with instructions on creating and anchoring a resource state, with the specific instruction that she was to offer this to her client as a means to refresh her system while they took a break from working on the problem. Together they created a series of powerful anchors for resource states including peace, love and joy. The client was given the responsibility to use and practice the anchors at home and even to develop several new ones over the course of several weeks. During that relatively short period, the woman stopped drinking and reported that she was feeling better than she had ever felt in her life.

This exercise orients the client towards positive resources. It sets up a present time experience of powerful internal and self-generated states. It begins by challenging the participant to choose an intensely pleasurable state. Whatever the state chosen, the experience of enhancing the state illustrates that all emotions are subject to conscious manipulation. As the state is enhanced, the actual feeling is abstracted from the original memory context. That is, as the feeling increases in intensity the memory fades away. The feeling tends to be transformed from something that *happened* to them to something that they *can do*. Repeated access to the memory provides practice effects for positive feelings. There is usually a surprising experience of memory enhancement.

This exercise enlists the participant in a series of pleasurable experiences which, superficially, have no relationship to drugs or treatment. The most important lesson here is that people can choose to feel better and there are simple techniques available to make that possible. This technique works best if the client thinks that it is unrelated to the problem. It should be something that '… they can just do for themselves.'

The presuppositions that underlie NLP's wholeness perspective are that people are fundamentally not broken and that every person has the resources necessary to accomplish their goals. This exercise orients the individual towards that wholeness and supports their capacity to awaken unused resources. It presupposes that people have access to memories that can be used as behavioral resources in the present (Andreas C. & Andreas, S, 1989; Andreas S, & Andreas, C., 1987; Bandler & Grinder, 1975, 1979; Bodenhamer & Hall, 1998; Cade & O'Hanlon, 1993; Dilts, et al., 1980; Erickson, 1954; Gray, 1997, 2001; Linden & Perutz 1998).

The exercise assumes Miller's (1956) discovery that the working memory store (short term memory) has a limited capacity. By emphasizing more and more features of the felt experience of the memory, we gradually abstract a feeling tone from the memory and allow the memory content to fade away. The "magic number seven plus or minus two" suggests that as more and more features of the feeling itself are emphasized, the content and context of the memory drop away. In this exercise, success is measured by access to a point where the state is no longer identified with a memory or memory context, but the client floats freely in a tranquil nether land associated only with awareness of the feeling and their capacity to control it.

The simple act of choosing a memory and manipulating the memory provides a powerful experience of self-efficacy. When used to frame outcomes, this serves the value of transcendence, the ability to out frame negative behaviors because of the strength of a positive outcome (Baumeister & Heatherton, 1995; Prochaska, 1994).

Invite the participant to choose an experience that made them feel wonderful. It may have been empowering, fulfilling, fun or ecstatic. It may be useful to find an experience of focused attention, love, competence, or stability. In choosing the state it is important to emphasize the following criteria for choosing a target state:

- Choose one specific moment in time (not a series of times). This might be experienced as a short movie or still picture, ending at the most intense part of the experience.
- The memory should be emotionally clean. It should not (intrinsically) carry the emotional baggage of regret or bad circumstances.
- The memory or circumstance should be stable over time and not subject to transformation (such as focusing on a present job or relationship that could be lost or destroyed).
- All examples should be experienced for themselves, without regrets or negative baggage. If a state cannot be used without self-pity or remorse, another state must be used.
- The example should have nothing to do with the problem state. Insofar as possible, this exercise would be presented as a prelude to or a break from treatment.
- Suggest that childhood memories of innocent experiences are just fine. Remind the client just to go for the memory in isolation. For any resource, the client should be encouraged to access it for its own sake.

As the client accesses the state, she should be asked to notice the difference between associated and dissociated experience–whether they are in the picture or out of the picture--and have them vary the intensity— bring it closer, make it brighter, make it louder. After each change ask the client to note how their experience changes. Each instruction is designed to provide a felt change in the experience and to provide practice in the manipulation of feeling by changing the submodality qualities of the experience (Andreas, S, & Andreas, C., 1987; Andreas C. & Andreas, S, 1989; Bandler & Grinder, 1975, 1978; Bandler & MacDonald, 1987; Bodenhamer & Hall, 1998; Dilts, et al., 1980; Gray, 2001; Linden & Perutz, 1998).

It is important to emphasize that not every remembered experience will have the impact of a photographic memory. Initial experiences are often weak and must be enhanced. This is the specific value of the submodality manipulations; a systematic means for controlling the valence and intensity of subjective experience. At the outset, whatever sense of the memory is available will work well.

Have the participant close her eyes and experience the memory. Let them note just how they get to the memory: what they notice first, a picture, a smell, a feeling? What comes next and next and next? One client described his access to a time of focused attention as first hearing confused sounds, then having a feeling of butterflies in his stomach. This was followed by a sharp smell of specific odors associated with the incident and another increase in feeling. He next found himself focusing on the face of someone and a further intensification of the feeling.

Advise the participants that your suggestions are just that— suggestions that they can try. If there is no picture at first, turn to the sound. If there is only feeling, stay with it and don't worry about the other parts. Reassure the participants that whatever sensory manipulation that they can use to enhance the feeling is just right.

In one of our groups, a color-blind participant asked what he should do with the instruction to turn up the vividness of the colors. He was advised to notice what would happen if he could. He reported an immediate increase in the intensity of the experience.

Remind your client to focus more and more on the qualities of the felt state. Overload short term memory with impossible dimensions of feeling: location, texture, depth, breadth, height, temperature, imagined color and imagined sound. As the she focuses on more and more of these, the context and content will be crowded out of working memory and she will be left in a powerful, peaceful ecstasy that carries the flavor and physical tone of the original state. It is a generalized state of autonomic arousal that is framed by the original state.

An important part of the exercise is the abstraction of the feeling from the memory. We begin with a remembered experience to gain access to a feeling state. We enhance the memory to increase the felt sense of the experience. We then focus more and more on the feeling in order to lose the connection with the memory and discover the feeling as something associated with the participant's own capacity to feel; independent of external influences. By abstracting it, we gain a completely transferable resource. By making it strongly pleasurable, we gain a motivation for practice, increased probability of use and a set of positive experiences that can compete with cravings.

At the end of the session, provide the client with instructions and an opportunity to practice the enhancement techniques with several more experiences of their own choosing. In practice, review and re-access the states at the beginning of your next meeting. This will begin the session with a positive bias towards the facilitator and more generally towards the techniques.

Client instructions

Think of a time when you were in love, or perhaps you were loving a pet or other small creature. Perhaps there was a time when you felt particularly empowered or free. You might think of having fun as a child or an early crush. You might think of your first dog or cat, or an experience of special competence. For now, choose a positive memory. It does not have to be the best thing that ever happened, just something that you'd like to enjoy again. Make it something that is complete in itself—something that will always be special.

Edit the memory so that you are focusing upon a repeating image or sequence that remains positive. Keep repeating the very best part.

A single memory is usually best. Focus on the best ten seconds of that memory. Gently turn your attention towards that one part. If your attention waivers, that's OK; gently turn your attention back to the very best part.

Think of a time when you felt wonderful.

Notice whether, in your imagination, you are experiencing the memory from within your own body, or experiencing it from outside like a movie.

If your memory seems to be just in your head, imagine that you can step all of the way into it. As you experience the memory, you may even notice flashes that feel like really being there, gently turn your attention to these. Take a few minutes to make sure that you are actually in the experience. When you begin to have the sense of really being there, even if it was only for flashes, come fully back into the present context.

Now that you have a sense of what it's like to relive the memory from within, step all the way into it and get a feel for it. Notice that you can step right into one of those parts where it all came alive. Step right into it. Notice what you are seeing and feeling and hearing. Notice the patterns of tension in your muscles. Notice who else is there in the memory and how you feel emotionally. Take a few minutes to get really familiar with the feel of being there. Enjoy it. Come fully back into the present.

Step back into the memory. Again notice how you can zoom right into the best part. As you do so, make believe that the memory is huge, bigger than life. Become aware of the sound and the directions from which the sounds come. Notice how the sounds enhance the experience. Come fully back into the present.

Now, return to the memory once more. As you do, notice that you can zoom right to point where you left off the last time; right to the very most intense part. Make it bigger and brighter and closer. Turn up the volume of the sound until the volume is just right for intensifying the feeling. Notice the rush of feelings and sensations. Pay attention to the feelings and notice where in your body the feeling starts and how it spreads through your body to peak intensity. Shake out the feeling and return to the present.

Return to the memory and zoom right back to the very best part. Turn up the brightness, bring it closer and turn up the volume on the sound. While you do these things, note the path of the energy through your body. As you notice the feeling getting stronger, begin to notice how the feeling moves. Notice whether it moves like a bicycle wheel or like a turntable. Does it move clockwise or counter clockwise? Notice that it moves further, faster and more powerfully. Notice whether it hums, what color it may have, whether it gives off sparks, glows or pulsates. As you do this, you will notice that the memory fades flickers and then goes away. That is just what we want. Let the memory go and focus on the feeling.

Continue to recycle the energy. Do it faster and faster until you lose any sense of the memory and find yourself floating, immersed in the feeling alone.

When the memory itself fades, but the feeling remains, you have crossed a subtle threshold. Emotion has begun to be something that you can do, not just something that happens to you. You have chosen to feel something and you now have subjective tools for doing it again. You can do it with any feeling that you have ever had.

Step back into the feeling. Do it quickly and notice that you can control it. Notice that control comes by gently turning your attention and **resting down** into the best part. The more you rest, the more you **gently** turn your attention **back** to a feature, the more control you will have. Take some time to discover how deeply you can enter the experience of pure feeling.

Anchoring the resource

One of the crucial insights of Grinder and Bandler (1995, 1979) was that otherwise brilliant interventions often have limited impact because their effectiveness is limited to the office or learning context. Because of this they fade to insignificance in the very places where they would be most useful. To prevent this, and to provide continuing access to a feeling of empowerment about the feelings created, we anchor the response.

Client Instructions

Start by accessing the state you've just been working on. Do it several times. Do it until the state arises quickly and you are conscious of a rush of positive feeling. Do each repetition as fast as you can, and find out what pace allows you the most enjoyment.

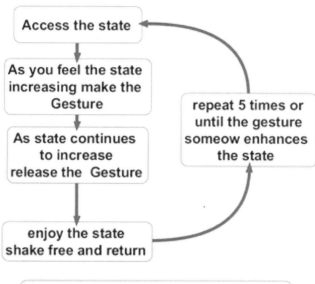

Self-Anchoring Process

Work with the state until it is content free, so that you can go right to the feeling. Step all the way into the feeling and immediately focus on the movement and the temperature and the texture of the feelings. Feel the rush. Enjoy it, spin it up, then shake out the state, and come fully back into the present.

Return to the state. As you do, notice that you can zoom right to point where you left off the last time; right to the very most intense part. Notice the rush of feelings and sensations. Enjoy them for a moment and then, return fully to the present.

Now that you have a real sense of how quickly and powerfully the state can come on, begin the anchoring procedure. Use a simple gesture, like touching the tip of your thumb to the tip of your pointer finger. *The first few times that you use the anchor, NOTHING will happen. Just doing the anchor will seem to get in the way.* After the third or fourth repetition, you will begin to notice that something is happening, and this can be very dramatic. Read through the next several paragraphs before you continue, then just do it.

1. Close your eyes and zoom right back to the most intense experience of the state.
2. As you experience rushing into the state, make the gesture.
3. Hold the gesture for about two seconds—*while the feelings are still increasing.*
4. Release your fingers, but keep your attention on the state.
5. Enjoy the state for another second or so.
6. Shake out the state (shake your body) and return to the present.
7. Repeat this sequence five to seven times, <u>or</u> until you really begin to notice a change in the experience whenever you make the gesture.

Once you have the clear sense that the gesture is adding to the power or depth of the experience, make the following change:

As you notice the change in feeling after making the gesture, quickly break and remake the gesture.

- Remake the gesture and hold it until you become aware of a new rush of experience.
- As soon as you begin to feel a positive change in the feeling, break and remake the gesture again.
- Repeat this pumping action until the experience becomes pleasurably intense.
- Shake out the state (shake your body) and return to the present.

For most people, pumping the gesture might mean gently rubbing the fingers together or it may mean gently pulsing the muscles while holding the gesture. I generally find that once the anchor has been created, pulsing the gesture works best. Find a method that works for you.

Take a little time and play with this anchor. Find out how you can intensify the feeling. Find out how you can change the timing of the gesture to make it work better. Find out how good you can feel. After a few minutes, come all the way out, and start over by accessing the state, making the gesture and pumping it.

After you've spent some time playing with the anchor, it's time to test your work. Up to now, we've depended largely on accessing the state directly, now we are going to try the anchor alone.

Test the anchor

- Clear your mind.
- Sit or lie comfortably and make the gesture.

- Notice any feeling that comes as you make the gesture
- Begin to pump the gesture repeatedly.
- Do your best to make the gesture at the first hint of a bodily feeling.
- Repeat the pumping action as you focus on the best parts of the experience.
- Enjoy the growing intensity of feeling.

With each pump, allow your attention to discover something better or deeper in the feeling. As you do this, enjoy more and more aspects of the feeling itself. Let your attention move fully into the feeling. Keep pumping until you have an intense experience of pure feeling. Shake out the state (shake your body) and return to the present.

As you begin the anchoring process, practice accessing the state very quickly so that you can be sure to leave behind any pictures, sounds or other contextual information from the original memory. If any of that lingers, time the pumping of the anchor so that it catches the first hint of feeling, before the pictures or sounds have any chance of appearing. You will find that the anchor will take care of it automatically.

The anchor stimulus or gesture should not be distracting. If you design or choose a gesture, choose one that takes minimal effort. As you make the gesture, relax your hand comfortably and make the gesture gently. Concentrate on the state **not the gesture**. Getting the state right is more important than getting the gesture. The important quality of the gesture is its consistency. Whatever gesture you use, do it the same way each time; do it quickly and easily.

Once you've created an anchor, you've really created a control button for the state. Here are some things to try. 1. Vary the intensity of the state by speeding up or slowing down your gestures—pump faster or slower. 2. Explore the feeling landscape that you have created and when you find a part that is particularly interesting begin to pump the gesture a little faster. 3. If you find yourself at a plateau, try stopping the gesture until you float back into a new and more accessible pathway, or simply turn your attention to another part of the landscape and pump faster.

Anchoring now provides you with a tool that, combined with simply turning your attention to some facet of the experience, will allow you to fully control the depth, scope and intensity of the state.

Chapter Ten

Pseudo-orientations in time

In the last chapter we covered some techniques that could be used with persons who are motivated to change but need some help. The help may have had to do with becoming conscious of an outcome, getting permission to pursue an outcome or finding the resources that allow new decisions to be made and new directions to be taken.

Here, we continue with our discussion of techniques that may be useful with motivated changers. We focus on using the well-formedness criteria for outcomes as an application of pseudo-orientations in time and follow with a brief discussion of the miracle question.

Working through a well-formed outcome

In general, we don't think of the well-formedness conditions for outcomes as a strategy for change. They represent, nonetheless, a powerful technique that takes advantage of Prochaska's (1994) strong principle of change and Erickson's Pseudo-Orientations in Time (1954) to stimulate change in many people.

We have already discussed the motivating power of Prochaska's strong principle of change. We have also linked it to Baumeister and Heatherton's (1996) idea of transcendence; that a strong motivator allows the actor to move past distractions.

Pseudo-orientation in time is a hypnotic technique pioneered by Milton Erickson and popularized significantly by Scott D. Miller and Insoo Kim Berg in the *Miracle Method* (Erickson, 1954; Miller and Berg, 1995). It appears in the literature of Neuro-Linguistic Programming (NLP) in various forms including the Smart Outcome Generator and timeline interventions (Andreas and Andreas, 1987, 1989; Bodenhamer and Hall, 1998; James and Woodsmall, 1988; Linden, 1998). It represents a significant addition to any clinician's toolbox and is not limited to hypnotic contexts.

A pseudo-orientation in time is an exercise in which the individual projects him/herself into a desired future for therapeutic purposes. It can be used to clarify goals and outcomes, to create motivation for change, to eliminate resistance to change and to elucidate the path to desired goals.

Here is Erickson's description:

> This technique was formulated by a utilization of those common experiences and understandings embraced in the general appreciation that practice leads to perfection. That action once initiated tends to continue, and that deeds are the offspring of hope and expectancy. These ideas are utilized to create a therapy situation in which the patient could respond

effectively psychologically to desired therapeutic goals as actualities already achieved. (1954, p.396)

Erickson suggests that the power of the technique lies in the client's experience of change as a pre-existent fact and its reliance on the unconscious mind's ability to create a future that incorporates his hopes and dreams. The pseudo-orientation in time is used "... to create a therapy situation in which the patient could respond effectively psychologically to desired therapeutic goals as actualities already achieved."

> "This was done ... using, a technique of orientation into the future. Thus the patient was able to achieve a detached, dissociated, objective and yet subjective view of what he believed at the moment he had already accomplished, without awareness that those accomplishments were the expression in fantasy of his hopes and desires (1954, p.396)"

The essence of the technique is the assumption that we can actively participate in the process of creating futures by accessing creative possibilities that already dwell within us. This process, of awakening unconscious resources to build creative futures, was named pseudo-orientation in time (Rossi, 1986).

Erickson emphasized that the technique takes advantage not of conscious fantasies which emerge fully formed and dismissible as mere imaginings, but on unconscious desires that represent preexistent goals and directions.

> *Unconscious fantasies* ... are not accomplishments complete in themselves, nor are they apart from reality. Rather, they are psychological constructs in various degrees of

formulation, for which the unconscious stands ready, or is actually awaiting an opportunity, to make a part of reality. They are not significant merely of *wishful desire* but rather of *actual intention* at the opportune time (1954 P. 421).

Erickson tells us that imagined future outcomes, are not just random thoughts or pipe dreams; they are expression of real desires and intentions. Such outcomes are built of fragmentary hopes that are just waiting for the right time to come together into realities. So, when we help someone to bring together a well formed outcome, we are helping them to formulate something that they already want and would do if they thought that they could.

When used as a change technique, per se, the criteria for well-formed outcomes are sometimes referred to as the Smart Outcome Generator. It is regularly represented in the literature of NLP. (Andreas and Andreas, 1989, Bodenhamer and Hall, 1988; Linden, 1998)

As previously noted, the procedure sets up 4 root criteria as follows: 1) Positive outcome; the goal must be stated as something that is wanted or a goal that can be attained. 2) The goal must be under the client's personal control; it should be active, not passive. 3) Sensory richness, concreteness; the goal must be represented by a concretely verifiable objective act or artifact. 4) Ecology-real life consequences; how will this change affect your life and the lives of the people around you (Andreas and Andreas, 1989; Bodenhamer and Hall, 1988; Cade and O'Hanlon, 1993; Miller and Berg, 1996; Linden, 1998)?

As the criteria are applied, the outcome becomes more and more real in the client's experience. Each step through the exercise introduces another level of complexity, another level at which the target outcome is integrated into the client's reality. Each step produces another round of practice, of trying on the target behavior in "a detached, dissociated, objective and yet subjective..." manner (Erickson, 1954, p. 396).

Where the technique departs from its standard use—as applied here—is in the source of the outcome. As originally formulated by Erickson, the outcome was to be generated unconsciously, assembling partially realized impulses towards action. In this case, we begin with a resource. Using the techniques from the last chapter for anchoring a content-free resource, use the same techniques for identifying an outcome based on the feeling alone. Allow the feeling or the part responsible for the feeling, to provide a vision of a future rooted in that feeling.

The exercise begins with the elicitation and enhancement of a powerful anchored affect, as generated in the last exercise. This forms an intuitive direction for choosing an appropriate outcome. Suggest that the client use that feeling to guide him to a future which would allow him to feel that way every day. What would they need to be doing? What would they need to add to their present experience to make this feeling the characteristic feeling of everyday life?

Have your client continue enhancing the state and begin to imagine the goals or outcomes that naturally come to mind and that resonate strongly with this state. Let him write a few examples down on paper.

The first criterion, that the goal must be stated in the positive is not immediately apparent to most people. We have all set goals that were stated as negatives: "I don't want to smoke." "I don't want to do X any more". The logic of the requirement, however, becomes immediately apparent. How many ways are there to *not* do something?

Fundamentally, the possibilities are endless. You can, *not* do something by doing anything else. A negative goal does not imply any direction. By contrast, a positive goal states, and states explicitly, a positive direction. "I want to have choices about when and if I smoke..." is a very clear statement. It provides direction and focuses the attention. I don't want to smoke anymore says nothing about direction.

The second criterion, the goal must be under your personal control, is a little more obvious. If my goal or outcome depends upon someone other than myself, it is by definition, invalid. "I want to win the lottery" is an invalid goal. "I want Joey to fall in love with me." is likewise invalid on its face. On the other hand, I may make a valid goal that will qualify me for Joey's attention or even develop a strategy that may enhance my chances of winning the lottery. Thus, "I would like to learn how to attract Joey..." is a valid goal.

Implicit in the second criterion is the concept of chunking, or manageability. The old saw asks, "How do you eat a whole watermelon?" The answer is, of course, "One piece at a time." When we think of a task, it is important not to bite off more than we can chew. A crucial element in whether a task is achievable is its size. If a task is too big, it will be abandoned. If the task is too small, It may seem inconsequential and not worth the time. Larger tasks can be broken up into sub goals. Smaller tasks can be chained to reach a larger outcome. In general, each individual has a preferred task size that must be considered (Andreas & Andreas, 1987, 1989; Bandler & Grinder, 1975, 1979; Cade & O'Hanlon, 1993; Dilts, Delozier, Bandler & Grinder, 1980; Dilts & Delozier, 2000; Miller & Berg, 1995)

Mihaly Csikszentmihalyi (1990) describes motivation in the flow state as a balance between boredom and frustration. The right task size provides sufficient challenge to make the task interesting. It remains accessible though just out of reach.

The third criterion is sensory specificity. In order to program the brain with a task or outcome, there must be a clear, specific result. Generalities like love and caring are nice thoughts but unacceptable outcomes. To transform them into acceptable outcomes they must be described in detail. How will you know you have it when you have it? This is a crucial piece of the strategy. If I want to be more loving, then I must know specifically what that means. Will I give more to charity? If so, how much and how often? Does it mean that I will actually listen to my spouse? If so, how will I let them know that I am listening and what evidence will I have that I am doing it? What will I see, hear and feel? What will I be doing? How will they respond? (Bandler & Grinder, 1975, 1979; Dilts, Delozier, Bandler & Grinder, 1980; Dilts & Delozier, 2000)

Criteria four and five require the participant to imagine the outcome in the context of their life more generally. Who will it affect? When do I want this? When might I not want this? If I am sick, would it be appropriate? If a loved one is in trouble, how would that modify the plan?

While working with long term addicts, sobriety is so highly valued that the context of that sobriety is often missed. What happens when a person who has been the identified problem, blamed for everything or just ignored, becomes a functioning human being? How does that affect his place in the family and the family structure? What will it mean in terms of changed responsibilities? How will his roles need to be redefined and how will he find his own place in the family and community?

One of the important implications of this exercise is that we are looking to create a real, internal image of the goal state. With each step of the process we are making our picture of where we are going more real and more complete. By the time we have gone through the first three or four criteria, we will know for certain if we want this or if we don't (Andreas & Andreas, 1989; Bandura, 1997; Bodenhamer & Hall, 1988; Cade & O'Hanlon, 1993; Miller & Berg, 1996; Linden & Perutz, 1998).

As the outcome becomes more and more real, it becomes a more powerful motivator. By the time we reach the final stages of this process, we can step into the reality of having it and really experience what it means to us.

Part of the particular genius of this exercise in its various incarnations is its use of Milton Erickson's Pseudo-Orientation in Time. This technique builds a desired future reality and then, from the perspective of that reality, builds the strategies necessary to reach it. Our presupposition is that if we build a goal that is consistent with our inner direction, all of our personal resources will naturally strive to achieve it. Since the outcome is a natural extension of who we really are, there are already resources in place that can be assembled to move in the appropriate direction.

By using the fantasy of experiencing the future, we can creatively assemble the resources in imagination and "remember" how to realize the goal. Because we have built a realistic picture of the outcome in imagination, charged it with the positive feelings that we hope to experience through it, and "remembered" steps that have "already" gotten us there, that future can be perceived as familiar, accessible and a reasonable hope (Bandura, 1997; Erickson, 1954).

As you come to the sixth criterion, encourage your client to step all of the way into the imagined future. Remind her to use her submodality skills to create the anticipated future. Make it bigger and brighter. Turn on the sound and notice where the sound is coming from. How does it feel to be there and where, specifically, does she experience that feeling? Encourage her to imagine that reality with the same kind of comfort that characterizes everyday experience. Where is that sense of familiarity located? What kinds of tensions accompany this new reality? What is her posture? What is she wearing?

As your client enjoys the present tense reality of their future outcome, have them casually begin to remember just how they got there. Let them begin to "remember" significant milestones and tasks on the way to that success. Have them begin with the last step that made it possible. What was the step before that? Get a list of four or five steps.

Make sure that the steps are expressed in the past tense. Encourage the participant to so identify with the future that the imagined past is expressed appropriately in the past tense.

Demand that the steps be concrete behaviors or outcomes. Fuzzy feelings will not do. Look for specifics. Things like "Call lenders and get rates and terms"… "Begin tonight by spending at least an hour with my children"… "Finish my GED".… "Enroll in the local community college for this specific program". These are all possible and appropriately concrete steps.

The concrete nature of the five steps is of crucial importance. Real futures have real antecedents. They have concrete foundations that can be identified as specific actions in the real world.

Research into the impact of imagined movement suggests that while the goal may be identified through the feeling state, future performance is only enhanced by the imagined performance of the process (Driskell, Copper & Moran, 1994; Martin & Hall, 1995; Pham & Taylor, 1999; Wohldmann, Healy & Bourne, 2007). If the steps are expressed as vague emotional states, demand from the participant how they can express each one in a tangible manner. Have them specify three ways that they will know that they are expressing it appropriately and how it will be connected to the goal.

By working backwards from the solution state, all of the steps become directly associated with the end result. Each one becomes a stepping stone to the larger end and so shares in its meaning and value. This is similar to John Overdurf's (2008) end-state energy. Context is crucial in order to maintain momentum. If one loses sight of the larger goal, the intermediate outcomes can become irrelevant.

Client Instructions

Use the following procedure to create and test your outcomes.

1. Is it stated in the positive, or can it be stated in the positive?

Don't think about what you don't want. A goal has to be stated as a positive thing, something you can hold in your hand or put in a wheelbarrow. I want to get my AA is a good goal. You can see yourself at graduation. "I want more choices about smoking" is a good goal. You can see yourself holding the pack, throwing it away, putting it aside, etc. I want to stop smoking is a non-goal. How do you see not-doing something? If you've started with something negative-"I don't want to do *this* anymore." or "I want to stop doing *this*."— find a new, positive way to see it or state it.

Positive goals have qualities that can be imagined. They can be seen, moved towards, and manipulated. They provide a focus for attention. Negative goals are much more diffuse. As their focus is negative, they can lead anywhere so long as it is away from the object.

2. Is it under your personal control?

A proper goal must be under your control. It must be something that you can do: saving money to buy a house or business, getting the training and connections to make a career change, finding new ways to change the way I think or behave. All of these are good goals. These are all things that you could do. If you could find the means, all of these things are choices that would be under your control.

Personal control also includes reasonableness. Is the goal you want realistic, or should it be broken up into sub goals or outcomes?

It is not uncommon for people to set goals or outcomes that depend upon some *deus ex machina*- winning the lottery, being rich, having fame- all are inappropriate both through their lack of specificity and through their dependence upon external agency. A motivating goal must be doable by the client. It must be specific enough to be manageable and within the capacity of the individual to complete on his/her own.

3. Can you specify three different ways in which you will know that you've gotten it if you get it?

How will you look when you have it? How will you feel? Who will be there? What will you see and feel? The more fully you can imagine getting it, the more powerfully motivated you will be to get it. Use all of your senses. Make believe that you are there and you can see it and feel it and touch it. It is really important in this step to really try to feel and see and hear and taste and smell your success. The more senses you can use the more motivating the goal will become.

If you're thinking about a business, think about the deed or licenses, hold them in your hand, feel yourself signing the deed, writing the menu, opening the front door on the first day of business. How will it smell? Can you smell the ink on the presses? Can you smell the food cooking in your ovens? Can you feel the hand shake from the boss and see his face as you are given the promotion? Can you see your family smiling with pride as you receive your license?

For this step, specify three of these kinds of things that will really let you know when it's yours.

Sensory specificity, concreteness, is a crucial part of realizing any goal. If there is no way to test that you've attained it, you do not have a well-formed outcome. An important part of this process is the idea of the TOTE process. It consists very simply, of setting up success criteria, (T) operating upon the environment to effect the indicated change(O); Testing for success (T) and either ending the operation(E) or continuing the process until the success criteria are met (Dilts, 1993, 1995). .

4. Do you want this all the time? Is it appropriate everywhere? Should it be limited to a specific context?

As we make a goal realistic, it is important to realize that it may not be appropriate at all times and places. Where do you want it, where don't you want it? Where would it be in the way?

Part of your job in making an outcome real is to discover if there are places where I don't want it. If my goal is to start jogging every day, will I expect myself to do it when I'm sick; in the middle of a blizzard? If I'm saving for a house, are there other things that might come up that may cause me to slow my pace? If I'm working on a degree, must I become a study-holic or are there times and places where I will not want to be locked down with my books. If I'm looking for a raise, will I still want it if the boss demands sexual favors or if I discover that he is breaking the law? If I'm buying a house or car and there is a question about its operability or title, will I still want it?

What are the problems that limit your goal? Use these to make it more real and specific.

5. What will it change in your life and in the lives of the people around you?

Real goals have real consequences. When you are sober, you will have new friends and new relationships. How will this work for you? Who will support you? Who will resist you? What will it cost in terms of relationships? What will it get you? Are you willing to take the chance? What will you gain in terms of new opportunities?

Despite classical addictions treatment, sobriety in not a valid goal or outcome. Insofar it is defined in terms of *not doing* certain behaviors, it is ill-formed. What do you really want that that is an essential part of? What do you really want from which that will blossom?

When you enter a new business or a new neighborhood, there are also changes in your life. How many hours will you have to work now? How many nights, how many weekends? Will you be better off or worse off?

Ecology, how change integrates with the rest of a person's life, is a crucial part of any therapeutic enterprise. By having the client consider the ecology of the sought-after goal, three things are accomplished: 1) The goal is integrated into the entirety of the client's experience. It is not simply a thing in itself but part of the fabric of existence. 2) By taking the time to assess the impact of the sought-after change, the outcome is further integrated and customized so as to meet the real world needs of the client. 3) The act of integrating and evaluating the outcome makes it more real in its consequences and more real as a potential motivator. Failure to include ecological considerations have destroyed otherwise powerful interventions (Grinder and Delozier, 1987).

One of the crucial events often triggered by just such an ecology check is the decision by the client that the outcome is for one reason or another ill-fitting or ill-advised. At this point, the following instruction is particularly apt. It can be important to emphasize here the root NLP presupposition that there is no such thing as failure, only feedback (Bandler and Grinder 1975, 1975a, 1979; Bodenhamer and Hall, 1998; Linden, 1997).

By this point, some people discover that the goal that they started with may be inappropriate. If this has happened do you, congratulations, you have made a crucial discovery about yourself and avoided one of life's major pitfalls. Take a few minutes to get centered and focus on that sense of really knowing what it feels like to have made this good decision.

You may have even discovered that there is something else that you would really like to do instead. If there is, work through the exercise again using this new outcome.

When you discover that you still want your outcome and that the exercise has made it more appealing than ever, continue with step 6.

At this point, having created all of the prerequisites for a powerful experience of the anticipated goal, the client steps into the future and begins to experience the future outcome. For most people doing the exercise, the careful processing of the steps provides more than enough information for the creation of a powerfully motivating experience.

As used in the Brooklyn Program, this step and the exercise more generally have been preceded by several weeks of systematic practice in visualization. Participants have developed significant expertise in changing the submodality structure of their experience and revivifying resource states. Here we ask them to use those same skills to create a sensory experience of their anticipated future

In practice, the participants are talked through these last steps as a group (or individual) visualization exercise. In this case, rather than using the quiet, ambiguous language of Ericksonian practice, the tone is more demanding and requests fast- paced responses to sensory based questions: What do you hear? From what direction does the sound come? Who is there with you? Who is not there? What does it smell like; take a deep breath and smell it.

The written instructions work well as they stand. The verbal cheerleading of an external prestige figure can provide a more powerful experience. Even the untutored reading of the given instructions to another participant can enhance the experience significantly.

6. Experience now, in your imagination, how you will look and feel, what you will see and hear when this is a reality.

What we are doing here is getting in touch with your future self. The self who has already accomplished your goals. It is important to feel and identify with this future you because he/she will show you how to get where you want to be.

Go back and get the image. See yourself in bright color. Experience the people around you. Think about how you will feel and how those around you will respond.

You can start with the three things that you used to let yourself know that you had it. Build from there and find yourself standing there, having it. Go through all of your senses.

Step all the way into it. See it, feel it and hear it from your own perspective. How do you feel having it? How do you hold yourself? Move into that same posture. What do you say to yourself? What do the people around you say?

Once you have a real picture of yourself having what you want, get into it. Enjoy it. Feel it and continue to feel it.

Here we have moved into a fully associated future experience. With this transition, we also are very careful to use the present tense. It is no longer about how the client will feel but about s/he feels NOW, already having it in their possession.

7. Move backwards from the final realization of the goal to discover the steps that make it possible.

Now, from that place, where you can see it and feel it and taste it and hear it, look back towards they when you did this exercise and find the steps that got you there. Begin with the last step, the finishing touch. Really be there. Ask yourself "What was the last step that I made that put it together?" Take your time. When you've answered that question, ask: "What was the step before that? And before that?" Take the time you need to find the steps that got you there. Keep the state, feeling like you've already got it. This will guide your mind to the steps that got you there.

Having had the future experience of a desired outcome, the client now adds the capacity to remember the steps that led to the success. In this situation the logic of getting there is best understood by moving backwards from the goal-state towards the problem state. If the client is successfully enjoying the goal, there are certain logical preconditions for that outcome. The last action or decision is the most accessible. Once that piece has been 'remembered', a logical yes-set is established for finding the other parts of the sequence. Moving backwards from the solution has the further advantage of linking all of the intermediate steps between the problem state and the solution state to the positive affect associated with the solution.

8. List the five steps necessary to get from here to there.

Think about the steps that you just learned from the future you. Break the list into 5 steps that you can handle. If necessary, the five steps can be five sub goals and you can do the process on each of the sub goals.

Here the participant writes down the five steps and the exercise is essentially over.

This intervention can have several outcomes. Some clients will respond by indicating that they know what to do and will begin doing it. They will immediately decrease the frequency of the problem behavior and over a relatively short period of time, with or without relapse, they will end the behavior,

In other cases, it will allow the client to identify the positive outcome and begin moving towards it, but they may still need further assistance in overcoming the problem behavior. They are now; however, well-motivated and other interventions can be used to good effect.

The structure of this technique aims to build a strong present time representation of a compelling outcome. In terms of Bechara's opponent process view of addiction, this exercise works in part by exercising the attentional faculties that addictions tend to disable (Baler & Volkow, 2006; Bechara, 2005; Chambers et al. 2007; Feil et al. 2010). As in the submodality and anchoring techniques from the last chapter, this exercise benefits from the effects of practiced attention and the use and enhancement of highly salient pleasurable states.

There is one more piece; this is about motivating change, not about reaching specific outcomes. Many clients who go through this process reach their outcome and find that it opens as a path of personal growth. This is the optimal result. For others, the process itself opens them to other possibilities that they couldn't have considered before. These too can become life paths. Holding to the presupposition that there is no failure, only feedback, as long as the exercise opens possibilities that can compete with the additive problem, it has succeeded.

THE MIRACLE QUESTION

Most simply, pseudo-orientations consist of sending an individual into another time or frame in which the problem at hand has been solved or the desired goal already obtained. Miller and Berg state the most basic formulation in their book, *The Miracle Method* (1995):

> Suppose tonight, after you go to bed and fall asleep, while you are sleeping a miracle happens. The miracle is that the problem or problems that you are struggling with are solved! Just like that! Since you are sleeping, however, you don't know that the miracle has happened. You sleep right through the whole event. When you wake up tomorrow morning, what would be some of the first things that you would notice that would be different and that would tell you that the miracle had happened and that your problem is solved? (Miller and Berg, 1995, p. 38).

The use of pseudo-orientations presupposes several things 1) The client already possesses the skills or abilities (resources) necessary to reach those goals. 2) Motivation often is established most powerfully from positive goals. 3) Imagined results can have the impact of actual experience. 4) Problems are generally not maintained in the same manner in which they were established. 5) Given the opportunity, clients will create meaningful futures rooted in their own capacities.

This approach assumes that each individual has within him or herself resources that are sufficient to solve the problem at hand or attain the goal sought. That these resources are not always apparent to the conscious mind is more often the problem than any presenting pathology.

The form of the question is designed, following Erickson's insistence on unconscious determinants, to avoid conscious interference with the process. It is, after all, *only* a fantasy. This frame allows the client to think the unthinkable. By avoiding conscious resistance, the client is freed to awaken the uncompleted urges and actions that constitute Erickson's idea of resources.

Resources are any experience or any memory of an experience that the individual has had. It may as likely be an imagined experience or a role play. Any or all can serve as a resource. The idea that people possess these kinds of resources was central to Erickson's approach and forms one of the basic presuppositions of NLP (Andreas & Andreas, 1989; Bodenhamer and Hall, 1998; Bandler and Grinder, 1979; Haley, 1973; James and Woodsmall, 1988; Linden & Perutz, 1997).

Erickson reflects the basic understanding of resources in the following: passage:

> *Hypnosis is not some mystical procedure, but rather a systematic utilization of experiential learnings -that is, the extensive learnings acquired through the process of living itself....* For example, mention may be made of hypnotic anesthesia or hypnotic amnesia, but these are no more than learnings of everyday living organized in an orderly, controlled and directed fashion. For example, nearly everyone has had the experience of losing a painful headache during a suspense movie without medication of any sort. Similarly, everyone has

developed an anesthesia for the sensation of shoes on the feet, glasses on the face, and a collar around the neck....

All of us have a tremendous number of these generally unrecognized psychological and somatic learnings and conditionings, and it is the intelligent use of these that constitutes an effectual use of hypnosis. (Erickson and Rossi, P. 224).

It was the significant contribution of Bandler and Grinder to make clear that these same capacities were available in non-hypnotic states (Bandler and Grinder 1975, 1975a, 1979; Bodenhamer and Hall, 1988; Dilts, 1993; 2001; Gray, 2011a.).

Often the resource remains unrecognized until the pseudo-orientation creates a specific resonance with it. In *The Miracle Method*, after going through the inventory of how things would be different—after the Miracle happened—Miller and Berg ask the client to think of experiences of similar non-problematic behaviors that correspond to the differences noted in the inventory from a time before they were patients. Pretreatment counterexamples are analyzed in terms of when, where and with whom the event occurred. These exceptions to the problem behavior are used to open the client to further use of those same resources and can be used to awaken the miracle in present experience. This resonance of past experience with anticipated futures is well supported by modern neuroscience showing that the vivid awakening of positive memories of past events enhances the capacity to visualize and plan meaningful futures (Freeman, 1998; Schacter & Addis, 2007a, 2007b).

Cade and O'Hanlon give it clear expression:

Central to the solution-focused approach is the certitude that, in a person's life, there are invariably exceptions to the behaviors, ideas, and interactions that are, or can be, associated with the problem. There are times when a difficult adolescent is *not* defiant, when a depressed person feels *less* sad, when a shy person is *able* to socialize, when an obsessive person is *able* to relax, when a troubled couple *resolves* rather than escalates conflict, when a bulimic *resists* the urge to binge, when a child does *not* have a tantrum when asked to go to bed, when an over responsible person *says* no, when a problem drinker *does* contain their drinking to within a sensible limit, etc. (Cade and O'Hanlon, 1993, p. 96).

Chambers et al. (2007) might consider the imagined outcome as a node in a neural network that might become a central element in an alternate life schema, separated from the problem-centered network. Insofar as it constellates, or brings together the elements of an already pre-existing sober network, it has the potential to awaken something akin to Miller's (2004) quantum change.

As we consider resources, we would do well to recall that most substance abusers regularly have long periods when they do not drink or drug problematically. All of them have periods of abstinence. Rather than being an expression of the *dry drunk syndrome*, they are here understood as experiences of already useful resources for future change.

One modification of this technique that may be very useful is be the identification of several of these pretreatment resource experiences, enhancing them using the submodality techniques described in the last chapter and anchoring those states. Those anchored states now become present time resources. Once anchored, those resources can be used as resources, counter motivations to urges, or enhanced to the point of meditative states. In any event, they can provide powerful examples of self-efficacy experiences with regard to the problem behavior and the more important context that made it unnecessary.

Chapter Eleven

The submodality blowout

NLP is rooted in the insight that all of what we do and experience is driven by internal representations of the world around us, mapped out in terms of the data of vision, audition, hearing, smelling and tasting. Those data, as perceptual and behavioral chains, become the schemas that drive behaviors and as the internal responses to external stimuli they become the representations of states. (Andreas, 2007; Bandler & Grinder 1975, 1979; Bostic St. Clair & Grinder, 2002; Dilts, 1985; Dilts, Bandler et al., 1980; Dilts & Delozier, 2000).

Beyond the simple chains of sensory experience that drive most behaviors, there is a vocabulary of submodalities, the details of sensory experience that represent how we feel about and evaluate the world around us. They determine meaning, including such dimensions as valence—approach/avoid, intensity, value as salience or importance, time relations and affective tone. All of the basic emotions are represented in terms of submodalities as are our responses to people, places and things. Significant among these are compulsions (Andreas, 2007; Andreas & Andreas, 1987; Bandler & MacDonald, 1987; Bandler, 1985, 1993; Bodenhamer & Hall, 1998; Dilts & Delozier, 2000; Gray, 2005, 20011b).

We have previously discussed compulsions as manifestations of salience hierarchies represented in the Orbito-Frontal Cortex. On a subjective level, these valuations are represented in the fine structure of perception: size, brightness, distance, volume, timbre, hue, saturation, movement, rhythm, warmth, etc. For each individual, the salience and desirability of any stimulus is marked out by submodality distinctions.

There are certain uniformities of representation that are general to people. Fuzzy and distant may give the illusion of temporal distance or unreality. Size and brightness and multi-dimensionality may give the impression of spiritual power. Glistening moistness with high foreground focus may signify desirability. Food and fashion stylists make their livings based on these kinds of generalities (Gray, 2011a).

In the world of addiction spectrum disorders, environmental cues and internal cues give rise to neural events that we identify as cravings. As part of this complex pattern of arousal, those same cues adjust our internal representations of people places, things and activities so as to make them irresistible.

Andreas tells us that when confronted with the cues that drive an addictive compulsion, the person experiencing the compulsion may be very aware of the felt desire, and even somewhat aware of the cues that have awakened them. They are often, however, not aware of the submodality dimensions of the internal representations that arise in response to the cues and actually drive the craving and create the feeling of compulsion.

One important facet of the submodality structure of any behavior or object seems to be that their placement in a biological context, a hierarchy of needs and values, provides them with boundaries which, if violated, change their absolute value. There is a limit on most things but that limit is not accessible to consciousness, it is a process driven limit. This appears to be closely related to the classical theory of behavioral extinction (Gray & Liotta, 2012)

In his research on submodalities, Richard Bandler discovered a way to drive submodality distinctions to such a point of intensity where they violate some undefined ecological boundary and become subjectively meaningless. He called this technique the Compulsion Blowout (Andreas, 2007; Andreas & Andreas, 1987; Bandler & MacDonald, 1987; Bandler, 1985, 1993).

The technique begins with the detailed comparison of two comparable objects. One of them is the object of a compulsive desire the other is not. For example, someone might have a compelling need to eat potato chips but not French fries. Because they are similar on many levels, these would serve as good exemplars.

Before making the comparison, care should be taken to note the physiological changes that accompany the report of a felt compulsion. Standard NLP practice requires that verbal reports of an inner state be confirmed by observation of external physiology. Note what happens to breathing, posture, voice tone, muscular tension, etc. and notice how specifically it differs from the non-compelled state.

After identifying the objects, the submodality structure of each is described in detail and then compared. This calls for the examination of things such as where do I perceive each in space? How near or how far are they from me? To what level is each focused or unfocussed, bright or dim, accompanied by sound or silent? What physical qualities do they have? Are they rough or smooth, warm or cold? As all of these distinctions about the objects accumulate, they have the net effect of producing a feeling of compulsion towards one but not towards the other.

After all of the differences have been elicited, each of those dimensions (only the ones that are associated with the increased experience of compulsivity) is tested by increasing or decreasing it to determine whether it will create an increase in the experience of compulsion for the previously non-compelling object. As the list of differing submodalities is manipulated, there should be at least one that makes a much more profound change in the feeling than any of the others. This is called the driving submodality, because it drives the feeling of compulsion.

Andreas makes the distinction here between two varieties of driving submodality. One varies over an infinite range. He notes that size in the visual channel is capable of infinite variation along a continuum from barely perceptible to unimaginably huge. If the driving submodality is of this variety, one very rapid expansion of the dimension to unimaginably intense levels is usually sufficient to extinguish its power to evoke the feeling.

Some submodalities vary through discreet ranges that give them specific meanings. Outside of those ranges they may have no meaning. Visual distance, in calibrating the fear responses is one such distinction. At one distance the object is irrelevant, at another, it evokes freezing, somewhat closer and it evokes escape behaviors; closer still, and it awakens fighting Blanchard, Blanchard, Takahashi, & Kelley, 1977)

In such cases, where meaning is delimited by a discreet range of submodality intensity, the submodality should be used to increase the feeling of compulsion rapidly and repeatedly, with very little time between trials. During the first several trials, the feeling of compulsion will increase but at some point, a subtle threshold is reached and the submodality will no longer awaken the compulsion. At the same time that the submodality ceases to work, the cue that originally awakened the compulsion will also stop working (Andreas, 2007).

We noted earlier that some techniques are appropriate for different levels of motivations and different varieties of the problem state. This technique deals with compulsion. In cases where a person is technically habituated (they can't stop the behavior but do not crave it or obsess about it) but not addicted (craving and obsessing about the behavior), this can easily end the problem.

The following outline is taken directly from Andreas (2007), Andreas, Steve. (2007, December). "Eliminating Unconscious Compulsions in Addictions" **The *Tenth International Congress on Ericksonian Approaches to Hypnosis and Psychotherapy*,** Phoenix, AZ.

Outline

Elicitation and Comparison. Elicit an experience of compulsion and a very *similar* experience of *not* being compulsed. (For instance, vanilla ice cream causes a feeling of compulsion, but vanilla yogurt does not.) Notice the observable nonverbal changes in the client in response to the experience of compulsion, so that you can determine nonverbally when the compulsion is gone.

2. Submodality Differences. Think of these two experiences *simultaneously*, and determine all the differences between the two experiences. (For instance, the ice cream is closer than the *yogurt.*

3. Testing Submodality Differences. Take *one* difference at a time, and vary it though a range, and find out how it changes the feeling of compulsion. (For instance, vary the distance of the ice cream from near to far, and monitor the experience of compulsion, both internally and externally.)

4. Find a "Driver" Submodality. Determine which of the submodalities is *most* powerful in changing the compulsion.

5. Infinite or Finite Range. Notice if the driver submodality varies through an infinite range or a finite range. (For instance, size of image can vary from zero to infinity, but distance may only vary from 3 feet to close to the nose.

6. Increase the Compulsion *Rapidly*.

    a. Infinite Range. *Very* rapidly increase the submodality to infinity (For instance, the size of the image of what compulses the client can be quickly increased to "larger than the size of the known universe.")

    b. Finite Range. Change the submodality rapidly through the finite ranges, and then repeats this over and over again, going in only *one* direction. For instance, the image is moved from 3 feet away to the tip of the nose, repeatedly, always starting at 3 feet—*not* yo-yoing back and forth. With either method, you should first observe a rapid increase in the compulsion, and then a decrease.

Testing. Pause for a minute or so, and then ask the client to think of the experience that previously elicited the compulsion, to find out if it still does. If the compulsion is still present, back up, gather information and find out what was missed. If the compulsion is gone, test to find out if it can be recreated in another modality, and if so, repeat this process in that modality.

Chapter Twelve

Changing the unwilling: The Brooklyn Program

Within the context of the criminal justice system, and clients who may have been referred for treatment, a large proportion of the client base is not interested in drug treatment and sees no reason to stop using drugs. Many of them occupy various levels of substance use disorders as noted in Chapter One. The group also includes persons with no discernable problem other than the mandate of the court. Needless to say this is a difficult population.

Most drug treatment programs take on the mission with a certain evangelical zeal: drugs are a disease and the treatment presented is the gospel. With great and often mistaken urgency they press upon their clients the need to realize that their lives are at stake.

Having observed the consistent failure of this approach, a decision was made to present what was designed to be a drug program and a self-improvement program that would satisfy the requirements for drug treatment but would never mention drugs. By providing useful skills and ecstatic experiences, we hoped to accomplish treatment without treatment. Because the program provided skills that were useful beyond problem contexts it was found to appeal to all kinds of audiences.

The Brooklyn program operated as an in-house substance use treatment program for the Federal Probation office in Brooklyn, New York during the period between 1997 and 2004. It began by treating offenders with verified histories of marijuana abuse or addiction and clients with no significant personal direction and expanded to cover offenders with all levels of substance use disorders. Participants met in a group format with one or two facilitators for two hours once a week over the course of the program's 16 week span. The program is fully manualized.

A mid-program statistical analysis of results from 100 recent clients (1999) found that those who completed the program did as well as other clients who had been referred for standard intensive outpatient treatment, but at a significant savings to the government in time and money ($3000 per successful client). The program is unique in that 1) It is non-confrontational and non-directive; the problem behaviors are for the most part never directly addressed. 2) It provides behavioral success criteria for each stage of the program so that facilitators can gauge participant performance. 3) Like coping strategy interventions, the program is focused on providing affective tools for enhancing choice and personal transformation; unlike more standard programs clients are never instructed as to where the tools *should* be used. 4) It assumes that the client base is unmotivated for treatment and is, for the most part there on an involuntary basis.

Participants reported significant increases in positive affect, and self-esteem as well as the government's savings in time and effort. Program completers were shown to have one-year abstinence rates of 29.6% as verified by random urinalysis (Gray, 2001, 2002).

Because it relies on the client to create the internal representation of the answer that he needs, its group organization belies its highly individualized and personalized structure. In effect, although the process is grossly the same for each client, its personal execution is transformed by the interpretation of the techniques used and the resources applied by each person.

The Brooklyn Program was designed to take advantage of depth psychological and humanistic hypotheses about human growth and development and their intersection with the Stages of Change Model set forth by James Prochaska and his colleagues (Gray, 1996, 2001, 2002, 2005, 2008; Prochaska, Norcross & DiClemente, 1994). More specifically, it was designed with the assumptions that:

- Substance use disorders are, in general, about the subjective utility of abused substances and behaviors and their capacity to produce an immediate but ultimately false sense of self efficacy (Gray, 2001, 2002, 2005, 2008; Zoja, 1990).
- The path to individuation / self-actualization represents a more salient, more personally rewarding set of experiences that are capable of out framing the addictive urge in the short term and creating meaningful future outcomes in the long term (Gray, 1996. 2001. 2002, 2005).
- In line with Prochaska's Strong Principle of Change, the identification of a more highly-valued future

outcome predicts movement from precontemplation to action in the stages of change model (Prochaska, 1994; Prochaska, Norcross & DiClemente, 1994).
- In line with Jungian assumptions about archetypal energies, meaningful, impactful future outcomes can be shaped by awakening a felt sense of personal identity—constellating the deep self—and using those felt experiences to create a set of outcomes that would meet Prochaska's requirement for a motivating future outcome (Edinger, 1971; Gray, 1996. 2001. 2002, 2005, 2008; Hillman, 1996; Prochaska, 1994; Prochaska, Norcross & DiClemente, 1994).
- In accordance with the work of Milton Erickson, later confirmed by Antonio Damasio, it was understood that present memories of past positive experiences could be used as resource states for acquiring the positive affect states that would drive the experiential base of the project (Erickson, 1954; Damasio, 1999).
- Because, according to James Hillman, any affective state experienced on a sufficiently deep level may be understood as archetypal, the affective states used to awaken the felt sense of self could be created and enhanced using simple conditioning procedures (Gray, 1996. 2001. 2002, 2005; Hillman, 1983).
- The experiences that drive the change could be created using simple behavioral techniques derived from the Neuro-Linguistic Programming (NLP) tool set (Gray, 1996. 2001. 2002, 2005; Dilts, Delozier & Delozier, 2000).

Although founded on presuppositions grounded in humanistic and depth psychologies, it soon became apparent that the principles upon which the program depended could be expressed on a deep level in terms of the structures in the midbrain dopamine system. Significant correlations were noted between the assumptions of the program and the behavior of individual dopamine neurons (Schultz, 2002), the instantiation of salience hierarchies in the Orbito-frontal cortex (a current review is presented in Kringelbach, 2005), and the differentiation between hedonic impact and incentive salience (Berridge & Robinson, 2003). Similar studies from the perspective of physiology provided a vertical integration of the hypotheses upon which the program was built and allowed for further refinements of the techniques employed.

The program begins by turning away from focusing on the problem and emphasizes that the participants can learn to enhance their memory, feel better emotionally, gain control over their emotions—choose how and when they want to feel differently, and finally, design a future that is meaningful to them. Beyond these outcomes, the only representation made to participants is that if they applied the techniques they would always leave the sessions feeling better than they did when they came in; if they didn't, it each session would be the most boring two hours of every week. Problems were deemphasized. In some cases the program was presented as laying a behavioral foundation for later work on the problem behaviors themselves.

In the first several sessions, participants are taught how to access and enhance a series of positive resource states using standard NLP submodality techniques. As any NLP practitioner knows, this submodality work begins with a striking enhancement of the remembered experience and so validates the first promise to clients that they will be taught memory enhancement techniques. During the same several sessions, the participants are taught to focus more and more on the feelings associated with the experience so that they discover a series of deeply-pleasurable transcendent states. These pseudo-meditative states are designed partly to provide feelings of self-efficacy, but also to provide powerful positive experiences that are strong enough to challenge the salience of the problem state.

Next, in sequence, the participants are taught to anchor several predefined states that they have accessed and enhanced during the preceding sessions. These include the experience of focused attention, a single good decision made in a systematic fashion, a moment of skill consolidation or streamlining of a learned behavior—riding a bike, driving a stick shift, an experience of pure fun or enjoyment, and an experience of confidence or personal competence. These resources are enhanced to ecstatic levels—to the point where there is virtually no shadow of the original content or context. Each state is anchored to a distinct hand gesture. The anchors serve three purposes:

- They make the resource transportable and accessible in multiple contexts,
- They create a relatively mechanical means for evoking and enhancing the anchored state,
- They create an automated access for later integration of these preliminary anchors into a more complex state (stacking anchors).

These five exemplars and the first level of stacked anchors were inspired by a set of anchors described by Carmine Baffa and were originally added to the program to facilitate the later exercises.

Once the anchors have been practiced and enhanced several times, participants are encouraged to practice them in multiple situations so that they generalize into other life contexts. This ensures that the new behaviors—access to the resource states—generalizes beyond the confines of the weekly session, A strong emphasis on homework and independent practice serves the same end. Participants are also encouraged to create several of their own anchors to make sure that they understand that all of this is under their personal control and that the resource states are theirs and theirs alone. A crucial element here is an emphasis on the development of efficacy tools and beliefs about the participants' own feelings (Bandura, 1997)

At about the seventh week, the anchors are stacked into a single anchor which has been labeled "NOW" and which, according to the author's understanding, creates a basic felt experience (constellation) of Jung's deep Self. This is important because it will provide an affective basis for creating a truly meaningful and compelling set of outcomes when in the last sessions we use the NLP well-formedness conditions to create a future that matches the function of the positive outcome in Prochaska's strong principle of change (Prochaska, 1994) and his observation that movement through the stages of change is propelled most significantly by the identification of a meaningful and compelling future.

The process continues with the collection and anchoring of another series of resources from various time periods in the participant's life. These consist of times when the participants felt good about themselves, things that they did well, things that they learned easily, meaningful jobs and roles that they held, and things they wanted to be when they were kids. These are again anchored, enhanced and integrated into the NOW anchor.

Finally, the felt state associated with NOW is used to create well-formed outcomes across several life domains: home life, occupation, spiritual life, relationships, intellectual life, and health practices. Each outcome is created by accessing the NOW anchor and imagining life in each of these domains through the affective window of the felt state "NOW". This results in future outcomes that are consistent with a deep, felt sense of personal identity. Superficial outcomes—wealth, sex, possessions etc. are discarded in favor of behavioral outcomes that characterize the kinds of behaviors that give expression to the constellated sense of the deep Self. The remaining exercises are devoted to enhancing the vision of the future and consolidating the learnings.

One of the more striking outcomes in the course of the program was the near universal and spontaneous use of the anchors for anger management. It seemed that as soon as the participants found out that they had a reliable means to control their emotions, they began to use the anchors to create choice about how they were feeling in the moment. This is all the more striking in light of our commitment to never tell the participants how or where to use the anchors.

In one case, an offender who had violated his several paroles for Bank Robbery on more than one occasion because of cocaine use came into the session and called the author aside. In a low voice he related that he had a problem. When encouraged to speak, he indicated that the previous night he had gone to his cop spot and found himself confused and did not know what to do. When asked what happened, he indicated that he just left. He was congratulated for his decision. After completing the program the offender completed his parole without incident and as far as can be ascertained has not returned.

During the regular weekly sessions, the anchors were used to provide access to ecstatic pseudo-meditative states. Outside of the treatment context, they often had the effect of bringing the subject out of depression or anger into a neutral state. One participant (who slipped past our attempts to screen-out psychiatric problems) suffered from bipolar disease. In the course of a trip home during the spring, her mother died and simultaneously she began to experience her depressive phase. When she returned from the trip she reported on her difficulties and was glad to say that she had not used any mind altering substances (confirmed by urinalysis). She indicated that she was disappointed in the anchors. She reported that when she found herself getting lost in depression, she fired off the anchors expecting a state of deep peace and meditative ecstasy. Instead, the anchors brought her up to a relatively positive neutral state that made the remainder of her time quite bearable.

Many of the participants indicated that they wished that they had had the program early in their correctional careers, whether inside a prison facility or while on the street, serving a term of community corrections, or better, before they had begun to offend.

The complete program manual is available as a hardcopy purchase or a free download at: http://www.lulu.com/shop/richard-gray/transforming-futures-the-brooklyn-program-facilitators-manualsecond-edition/ebook/product-17463559.html or write the author.

Chapter Thirteen

Postscript

Addiction spectrum disorders have been approached as a medical condition since the 1950s. On some level, this has been useful for obtaining funding and government support, on another it has befuddled our thinking about the problem and skewed investigation towards the biophysical end of the spectrum.

Because of the maps that guide medical research, we tend to look at these problems through the narrow scope of disease, progression and death, not adaptive responding by a nervous system seeking balance. The Jellineck model of a primary, progressive disease with biological, psychological and social elements, although sometimes true, is typically interpreted in a manner that obfuscates rather than clarifies (Ramskogler et al., 2001). Our map barely reflects the territory.

One of the most difficult implications of the medical perspective is the idea of agent. Diseases have agents. Infections have bacteria. The flu has a virus. Cancer may have chemical or viral precursors. Mad-cow disease has Prions. It is not surprising then that when we look at addictions, we need to find a causal agent. Our culture naturally points towards the substance or, alternatively, to the genetic makeup of the addict. In either case, we're told, if the drug wasn't there, the problem wouldn't appear.

There is, however, evidence that suggests that this is not necessarily so. In chapter three we reviewed three studies that showed fairly conclusively that addictions may be less about the substances than they are about the people who use them and the contexts in which they are used. In the rat park study we found that, even when addicted, rats who had the opportunity seemed to have more important things to do than to get high. The returning Vietnam Era soldiers had a 95% remission rate after months of heavy, daily heroin use, usually without treatment. Pain patients, without previous abuse histories, had an addiction rate of two-one-hundredths of one percent, despite persistent daily use. This suggests that on some level, the problem is not a specific property of the drugs themselves.

Moral panics have changed the face of our encounter with drugs. Many of them have been politically instigated; many of them remain as a matter of mental inertia. Heroin and Methadone are regularly used for pain control in other countries without problems. Every day, more medical uses for marijuana are being discovered. When it was legal to do so, there was a long history of studies that showed the beneficial effects of LSD. Luminaries such a Cary Grant and Bill Wilson—the founder of AA—are reported to have had the breakthroughs that allowed them to live free of addictive problems under its influence. MDMA and psilocybin are being tested for their capacity to ameliorate the pangs of death in terminal cancer patients and the Israeli Defense Force is testing marijuana as a treatment for PTSD...

The medical model has driven research into the physiology of addiction and has discovered, in its search to find pharmaceutical answers, that addiction uses the same circuits as normal, biological motivation. Attempts to medicalize these findings have led to addiction being defined as a brain disease that hijacks the normal mechanisms of motivation and emotion. On closer inspection, however, addiction blends seamlessly into compulsions that are not drug driven. Compulsive gambling, shopping addictions, sexual addictions and OCD all seem to participate in the same systems, often with the same levels of intensity. Love, conscientiousness, flow and various pleasures are rooted in the same mechanisms. Falling in love appears to be a dangerous addiction.

We must, however be thankful for the neurological studies of addictive process because they have not only uncovered the mechanisms of addiction and motivation, but they provide us with significant cues about how to think about these problems in a way that resonates with deep physiology as well as subjective experience.

The firing patterns of dopamine neurons reveal a pattern of biological preference that appears to hold true on every level of integration. Reinforcers that are surprising are deemed more valuable than those that are not. Those that increase in value are more important than those that remain stable. Stable reinforcers become habituated. Undependable reinforcers fall to the bottom of the hierarchy and from there fade away. B. F. Skinner is immortalized in neural firing patterns.

We have learned that within the midbrain dopamine system there are neurally programmed hierarchies of preference and value as well as hierarchies of fear and avoidance. The hierarchies are partly driven by the neural firing patterns above and partly by the sensory richness of their representation in the orbito-frontal cortex. The hierarchies rank those representations from intense and sensory specific parameters nearer the mid-brain to less intense and less motivating diffuse concepts at the frontal poles.

We know that drugs function by affecting the output or persistence of dopamine in these motivational systems. Artificially increasing their salience by direct chemical influence, the drugs, their cues and associations are felt to be more valuable, more important and more significant than they ought. Craving lies at the heart of addictive spectrum disorders.

More recent research has shown that the process of addiction leads to a down regulation of the critical and inhibitory patterns in the frontal lobes. Bechara's dual process pattern of addiction neurology tells us that beyond craving there are deficits in higher order inhibitory centers that must be dealt with.

One of the things that we often forget is that if our brains did not have these capacities anyway, drugs could not provide them. Richard Alpert, Timothy Leary's partner at Harvard University, tried LSD and decided that it was too important to just play with. So, he went to India to find a Guru who could teach him how to use it. Equipped with three massive doses of LSD, Alpert traveled until he was confronted by a wizened old sadhu who looked him in the eye and demanded, "Give me the medicine." Alpert complied and watched as the Guru took the drug. He then sat there watching the old man but the drug had no apparent effect. At some point, long after the first dose should have peaked, the old man looked up again and asked for a second hit. Alpert complied. Again, he waited and again there was no change. After the scene repeated for the third time that day, with the chemical apparently having no effect on the old teacher, he realized that this man had been more places in and out of his own head than drugs could ever match. LSD was irrelevant (Dass, 1971).

Rats in parks, GIs returning home and patients freed of pain; all had something more important to do with their heads than worry about drugs. Beyond the influence of drugs, there is usually something deeper, more relevant, more immediate and more complete than drugs and when we become aware of it, drugs become irrelevant. Even Alpert's Guru had better things to do.

Every day, people find a job, a relationship, a spiritual reality or an ecological niche where their lives are better, fuller and more meaningful than any drug induced illusion. For many others, for whom life holds out little hope, the drugs may be the best they have ever had. For many of these we can structure experiences that can change their lives.

The hierarchies that order our preferences are often context dependent and are themselves defined by superordinate values which give them meaning. Identity is defined by our place in the universe. Who I am, defines in turn what I believe and what is permitted to me. These beliefs and permissions, in their turn, shape the way I understand my own capacities; while those capacities shape the applications of behavior.

Problems may be out framed or transcended by changing the structure or meaning in the levels of integration that stand above them.

Awakening to new levels of meaning can change who we are, what we believe and what we can do.

In the context of addiction spectrum disorders, we can accomplish these kinds of changes by changing identities, beliefs and the larger definitions about who I must be. We can change them by awakening to flow states that spread through life as patterns of self-actualization and growth. We can awaken to quantum change or we can just do something different.

The possibilities for change are almost without limit. In some cases it may be as simple as taking a different subway or getting a new job. Sometimes it is a question of being encouraged to become what you've always known you needed to be. Some of our clients will need to sit down with some deep feelings and figure out what is really important to them and then begin to pursue it. Others still will need a tool that they can take with them; a button that can change the way they see the world or an internal voice that reminds them of just who they *really* are.

There are others for whom addiction spectrum disorders have really become a problem and they don't know it, or just aren't interested in help right now. Sometimes they need to be convinced that there are good things waiting, sometimes they need to be ambushed by good feelings and absurd good news.

There is something for everyone. It is our task to help them discover the map to get there and the tools they already have to awaken the treasures within.

The NLP Approach

Throughout this book we have made an effort to make distinctions: distinctions between different kinds of problems that have been labeled addiction, distinctions between the classical notions about addictions and their clinical reality, distinctions about the causes of 'addictions' and their current maintenance, and distinctions about what 'first things first' can mean. Using these distinctions we have provided a series of NLP-based interventions.

NLP does not deal with diagnostic categories but with presenting problems framed as positive, well-formed outcomes. The nominalization 'addiction' is too broad and imprecise to have any real clinical meaning within the NLP frame. As a result, our first task is to metamodel the problem state and to find out what the problem is and what the client wants positively.

To this end, we have already seen that behaviors labeled as addictions can be situational (constrained by time and place), only habitual (compulsive or pattern-bound), simple questions of bad judgment, actual lifelong problems or just labels. Sometimes addictive behaviors, as they are called, are related to shame, sorrow, pain, anger, depression, fear, and other internal states as contexts. When the contexts are gone, so is the addiction. In each case we are dealing with a very different phenomenon. *There is no single thing comprehended by the nominalization, addiction.*

Classical addictions treatments are remedial and typically framed as helping someone to stop doing something that has become a problem to the client or to someone else. This is by its nature ill-formed and not subject to a meaningful intervention. NLP seeks to create a generative answer to the client's request, something that allows them more choices than those offered by either standard addiction treatments or the 12 step model. 'How do you need to grow to make your life work more effectively for you?' is our frame for action and our directive for intervention; how can we help you to have more choices?

Classical interventions have tended to be over inclusive in their diagnoses, determining that addiction for all is a lifelong primary illness ending in abstinence or death. We know that this is not true except for a small minority of cases whose genetics and history have conformed them to this model. We also know that many people previously diagnosed as addicts have returned to a normal life-style that may include the use of alcohol and other drugs in moderation.

Insofar as so called addictions are rooted in the substances themselves, rather than some context or co-occurring pathology or problem state they can often be treated quickly and easily with the compulsion blow-out (for persons who are dependent but not craving for and obsessing about the substance or behavior), by helping them to find access to positive physiology, or by helping them to find and realize a life calling structured as a well-formed outcome.

NLP is more than a group of techniques. We have created techniques and patterns that work and work well for standard problems. The V/K-D—RTM protocol works wonderfully well to cure phobias and non-complex PTSD. The rapport pattern works to create powerful connections between people. The Spelling strategy works to improve spelling for almost anyone. The Brooklyn Program and its various elements have seen successful application in the realm of substance use disorders, where nothing else has seemed to work. Nevertheless, at the level of techniques we are not doing NLP we are using techniques derived from it.

At the heart of NLP is the modeling strategy: finding out how individuals structure their skills and problems so that those skills can be replicated and the problems alleviated. It is the active parsing of behavior into sequences of identifiable elements with specific relationships to each other and the world-out-there that makes NLP what it is.

Modeling is a recursive, hypothesis generating process in which the modeler frames a possible analysis and tests it with the client and their observed behaviors. The process repeats with continuing refinements of the model until a model of the skill or problem is created so that the skill can be replicated or the problem alleviated. The heart of this practice is summarized in the TOTE model and can be found as a more complete methodology and ground for theory in Grounded Theory.

Emerging from the author's own application of NLP to addictions and other problems, one of the crucial elements in change is the identification of a more valued outcome on a higher logical level. In NLP this may be understood as a hierarchy of criteria. From Dilts, we understand it in terms of his neurological levels. Michael Hall understands it in terms of Meta States and Connirae and Tamara Andreas have formulated it in terms of Core States. More generally, in the world of mainline psychology, it comes to us through the work of James Prochaska and his Strong Principle of Change. This indicates that most of the progress towards change in behavior, whatever the context, can be predicted by whether or not a person has identified an outcome that is more important to them than the problem behavior itself (Prochaska, 1994; Prochaska, Norcross & DiClemente, 1994).

Throughout this book and especially in the case illustrations that have been presented, we have pointed out that the client often knows exactly what they need. They know what is more important to them than the problem and, if they could get to it, it would allow them to live a normal life. The coke 'addicted' mafia lawyer needed to get back to church. The heroin addicted GI needed to get out of Vietnam and get back home. Sometimes they don't know what they need. The OTB gambler just needed permission to take a different route to work. The speed-balling woman, needed a higher value that required her attention, her baby, in order to stop even for a year. In each of these cases, the apparent addiction crumbled in the face of the right answer. These do not come through techniques or through diagnosis, they come from the basic procedures of NLP, Modeling, Rapport, the metamodel and persistence. Other tools are useful as are the presuppositions but there are no panaceas or magic bullets.

Assuming that the client wants help, the first thing to be done is to create rapport. This can be done with mirroring, matching and the rest but is best done by actually being interested in and caring about what the client has to say and the way they say it. In such cases we almost naturally mirror posture, breathing, tone and pace, and match the flow of sensory information by tracking predicates. But we may also want to remember the value of literally repeating the words that are most important to client. From here, the way forward is varied:

- What is your problem and what would you like to do about it?
- When does this happen or how do you know it's time?
- What needs to happen in your life in order for this not to be a problem again?
- Are there times when it was unnecessary, or you didn't even think about it? What were you doing then?

- Was there ever a time when this was not a problem for you? What were you doing then?
- What is the most pressing issue for you?
- Can you each me how to do that or to feel that way?

Many of these will inevitably evoke an ill-formed response that needs to be meta-modeled to generate a sufficiently specific problem or desired outcome. And this answer or set of answers will need to be transformed into a well-formed outcome using the smart outcome procedure:

- The outcome must be stated as a positive thing or experience; something wanted, not something unwanted or ended.
- The outcome must be something that is under the goal seeker's personal control which also implies that the task should not be stated too broadly.
- The outcome must be specified in terms of multiple levels of sensory experience; it must be described in terms of what can be seen, heard, felt, tasted or smelled.
- The outcome should be evaluated for ecology; what it will change in the person's life and the lives around them?
- The outcome should be imagined and experienced in fantasy as fully as possible (Andreas and Andreas, 1989; Bodenhamer and Hall, 1988; Cade and O'Hanlon, 1993; Dilts, Delozier & Delozier, 2000; Hallbom, 2014; Linden & Perutz, 1998).

In some cases your client will be able to tell you precisely what they need to do as the process uncovers a simple procedural answer:

permissions, reasserting a life calling, becoming involved; try that. If it reveals that the problem is susceptible to one of the patterns already discussed in this book, it might be time to test one of those. If that doesn't work, do something different. Probe more, ask more questions, find the structure of the behavior and the submodalities that drive it. Learn how to do the behavior yourself so that you can help to generate creative solutions.

In some cases the simple things work, in others they don't. Sometimes a generative restructuring of the individual, in terms of their own most deeply held values can create significant change.

Robert Dilts' Neurological Levels can be used to step people through all of the elements of the problem, from environmental constraints and reactive behaviors, to beliefs, values, identities and spiritual realities that can restructure the individual and out frame the problem behavior. Where and when does this happen and not happen? What are the specific behaviors and thoughts when it does or does not happen? What do these behaviors tell you about your capacities no matter whether the behavior is good or bad? What do all of them imply about what you can do? Having these capacities, independent of the way you've expressed them in the past, what do they tell you about what you believe and what you value? Or, having these capacities, what previously unconscious beliefs and values do they reveal? Believing and valuing these things, who are you on the deepest level? If you had to describe yourself in terms of a character from fiction, myth, religion or your own history, who would you really be? Name that person and character. Step all of the way into that identity and adopt the posture, facial expressions, manner of breathing and speaking that go along with it. Finally, as that person, standing before your source, your spiritual source, or your place in the universe, how do you feel and understand all of the rest? How are you empowered?

Adopt the posture and physiology of that state, align your entire being with that state, settle into it and be comfortable with it and take it with you as you revisit the levels in reverse with each one informed and transformed by this state.

The Brooklyn Program restructures people by providing them with deep experiences that are spontaneously recognized as being spiritual in nature. Stripped of content and context, magnified using submodalities, simple memories become gateways to powerful meditative states. When connected to a simple, personal, self-anchor, these states and the ability to make new choices that they imply, begin to positively impact the personal experience of the participant. Going one step further, using these deep states as vehicle to creating meaningful futures and applying the NLP well-formedness conditions for outcomes, creates a positive outcome, a sense of calling, a path for self-actualization and individuation that renders substance use irrelevant.

Connirae and Tamara Andreas' Core Transformation process follows a similar path by focusing on the positive intent of the behavior, whether positive or negative. Following that intent experientially into deeper and deeper layers of the psyche, the process leads to a spiritual level of experience that then is used to reframe each of the behaviors and felt states as the path of discovery is followed back to the original target behavior. The process changes the meaning of the behavior, its triggers and often the client's life.

In each of these generative change processes change arises from a deep encounter with levels of experience that are much more powerful, much more complete and much more natural to the client than any drug-induced experience could ever be.

There are also many times when clients are uninterested in change for their substance-related problems. Don't fight them; in NLP admitting that you have a problem is unnecessary, but doing something extraordinary often is. Such clients, such as court mandated clients, may, however, have other problems that they would like to work on. If you can work on these, you have the possibility of creating a deep state linked to deep criteria that can solve or move them into a desire to solve the problem for which they were referred. Moreover, in the process you have created a rapport, a therapeutic relationship that may support further work.

The Brooklyn Program was characterized by a group of clients (we had to refer to them as offenders) who, for the most part were uninterested in change. They were there because it was a condition of their probation. So we told them five things which though true, did not confront their resistance.

- We know that you are here for drug treatment. This will count as drug treatment, but we will never mention drugs
- We will teach you how to enhance your memory
- We will show you how to feel better emotionally, gain control over your emotions, and how to choose how and when they want to feel differently
- We will also show you how to design a future that is meaningful to you.
- Finally we promised that we will never tell them how and where they should use the skills learned in the program, unless they asked.

Although never mentioned, it was just these kinds of skills that would result in the kinds of controls and directions that would impact the problem. This however was a generative approach, not a remedial one, so the solution of the drug problem was a natural result of personal growth.

There is something for everyone. It is our task to help them discover the map to get there and the tools they already have to awaken the treasures within.

References

Alexander, B. L., Beyerstein, P. F., Hadaway, B. K. & Coambs, R. B. (1981). Effect of Early and Later Colony Housing on Oral Ingestion of Morphine in Rats. *Pharmacology, Biochemistry & Behavior*, Vol. 15, pp. 571-576: 1981.

American Psychiatric Association (APA). (1994). *Diagnostic and Statistical Manual of Mental Disorders. 4th ed.* Washington, DC: American Psychiatric Association.

Andreas, C. & Andreas, T. (1994). *Core Transformations*. Moab, Utah: Real People Press.

Andreas, C. (1995). *Core Transformation: A Profound Way to Let Your Inner Being Emerge.* Boulder, CO: Core Transformation International.

Andreas, C. (2002a). The Core Transformation Story: How the process came to be; Acknowledgements and History. Retrieved on 07/30/08 from **http://www.coretransformation.org/ct_story.htm**

Andreas, C. (2002b). Sample Case Study. Retrieved on 07/30/08 from http://www.coretransformation.org/sample_case.htm

Andreas, C., & Andreas, S. (1989). *The Heart of the Mind*. Moab, UT: Real People Press.

Andreas, S. & Andreas, C. (1987). *Change Your Mind— and Keep the Change*. Moab, UT: Real People Press.

Andreas, S. (2006). *Six Blind Elephants: understanding ourselves and others volume I: fundamental principles of scope and category*. Moab, UT: 2006 Real People Press

Andreas, S. (2007, December). "Eliminating Unconscious Compulsions in Addictions" The *Tenth International Congress on Ericksonian Approaches to Hypnosis and Psychotherapy*, Phoenix, AZ. Retrieved on 08/01/08 from http://www.erickson-foundation.org/10thCongress/HandoutCD/Presenter%20Handouts/Andreas/Compulsions.pdf

Asbell, H. C. (1983). Effects of reflection, probe, and predicate matching on perceived counselor characteristics (psychotherapy, interpersonal attraction, Neurolinguistic Programming (NLP)) (Doctoral Dissertation, University of Missouri at Kansas City, 1983). *Dissertation Abstracts International*, 44(11), 3515. Retrieved November 24, 2006 from http://www.nlp.de/cgi-bin/research/nlp-rdb.cgi?action=res_entries.

Austin, J. T. & Vancouver, J. B. (1996). Goal Constructs in Psychology: Structure, Process, and Content. *Psychological Bulletin*, 120(3), 338-375.

Baler, R. D. and N. D. Volkow (2006). "Drug addiction: the neurobiology of disrupted self-control." *Trends in Molecular Medicine,* 12(12): 559-566.

Bandler, R. (1985). *Using Your Brain for a Change*. Moab, UT: Real People Press.

Bandler, R. (1993). *Time for a Change*. Capitola, CA: Meta Publications.

Bandler, R. (1999). *Introduction to DHE*. Chicago (Audio).

Bandler, R. & Grinder, J. (1975). *The Structure of Magic I*. Cupertino, Calif.: Science and Behavior Books.

Bandler, Richard & Grinder, John. (1975a). *Patterns in the Hypnotic Techniques of Milton H. Erickson, MD, Volume 1*. Cupertino, CA: Meta Publications.

Bandler, R. & Grinder, J. (1979). *Frogs into Princes*. Moab, UT: Real People Press.

Bandler, R. & Grinder, J. (1982). *Reframing: Neuro-Linguistic Programming and the Transformation of Meaning*. Moab, UT: Real People Press.

Bandler, R. & MacDonald, W. (1987). *An Insider's Guide to Submodalities*. Moab, UT: Real People Press.

Bandura, A. (1997). *Self-Efficacy: The Exercise of Control*. NY: Freeman.

Bargh, J. A. (1997). Bypassing the will. In M. G. Shafto & P. Langley (Eds.), *Proceedings of the 19th Annual Conference of the Cognitive Science Society* (p. 852). Mahwah, NJ: Erlbaum.

Bankson, Michael G., & Yamamoto, Bryan K. (2004). Serotonin–GABA interactions modulate MDMA-induced mesolimbic dopamine release. *Journal of Neurochemistry, 91*(4), 852-859. doi: 10.1111/j.1471-4159.2004.02763.x

Bateson, G. (1972). *Steps Towards an Ecology of Mind*. New York: Ballantine.

Baumeister, R. F. & Heatherton, T. F. (1996). Self Regulation Failure: An overview. *Psychological Inquiry, 7*(1), 1-15.

Bechara, A., Damasio, H. & Damasio, A. R. (2000). Emotion, Decision Making and the Orbitofrontal Cortex. *Cerebral Cortex, 10*(3), 295-307

Bechara A. & Damasio, H. (2002). Decision-making and addiction (part I): Impaired activation of somatic states in substance dependent individuals when pondering decisions with negative future consequences. *Neuropsychologia, 40*(10), 1675-1689.

Bechara A; Dolan S; Hindes A. (2002). Decision-making and addiction (part II): Myopia for the future or hypersensitivity to reward? *Neuropsychologia*, 40(10), 1690-1705.

Bechara, A.; Damasio H.; Damasio, A. & Lee, G. (1999). Different Contributions of the Human Amygdala and Ventromedial Prefrontal Cortex. *The Journal of Neuroscience*, 19(13), 5473-5481.

Bechara, A. (2005). Decision making, impulse control and loss of willpower to resist drugs: a neurocognitive perspective. *Nature Neuroscience*, 8(11), 1458 – 1463.

Becker, H. (1963). *Outsiders: Studies in the Sociology of Deviance.* New York: The Free Press.

Berger, P., & Luckmann, T. (1967). *The Social Construction of Reality*. New York: Anchor Books.

Berridge, K. C. & Robinson, T. E. (1998). What is the role of dopamine in reward: hedonic impact, reward learning, or incentive salience? *Brain Research Brain Research Reviews,* 28:309–69.

Berridge, K. C. & Robinson, T. E. (2003). Parsing Reward. *Trends in Neuroscience*. 26(9), 507-513.

Bickel, W. K., B. P. Kowa, et al. (2006). "Understanding Addiction as a Pathology of Temporal Horizon." *The Behavioral Analyst Today,* **7**(1): 32-47.

Bickel, W. K., M. L. Miller, et al. (2007). "Behavioral and neuroeconomics of drug addiction: Competing neural systems and temporal discounting processes.*" Drug and Alcohol Dependence,* 90, Supplement 1(0): S85-S91.

Bickel, W. K., R. Yi, et al. (2008). "Cigarette smokers discount past and future rewards symmetrically and more than controls: Is discounting a measure of impulsivity?" *Drug and Alcohol Dependence***,** *96*(3): 256-262.

Bickel, W. K., R. Yi, et al. (2011). "Remember the Future: Working Memory Training Decreases Delay Discounting Among Stimulant Addicts." *Biological Psychiatry,* **69**(3): 260-265.

Blanchard, R. J., Blanchard, D. C., Takahashi, T., & Kelley, M. (1977). Attack and defensive behavior in the albino rat. *Animal Behaviour* 25: 622-634.

Bodenhamer, B. G. & Hall, L. M. (1997). *Figuring Out People - Design Engineering with Meta-Programs*. Williston, VT: Crown House Publishing.

Bodenhammer, B. G, & Hall, L. M. (1998). *The User's Manual for the Brain: The Complete Manual for Neuro-Linguistic Programming Practitioner Certification.* Institute of Neuro Semantics.

Bostic St Clair, C. & Grinder, J. (2002). *Whispering in the Wind*. Scotts Valley, CA: J & C Enterprises.

Bouton, M. E. & Moody, E. W. (2004). Memory processes in classical conditioning. *Neuroscience & Biobehavioral Reviews*, 28:7, 663-674.

Brockman, W. P. (1980). Empathy revisited: the effects of representational system matching on certain counseling process and outcome variables. (Doctoral Dissertation, College of William and Mary, 1980). *Dissertation Abstracts International*, 41(8), 3421. Retrieved November 24, 2006 from http://www.nlp.de/cgi-bin/research/nlp-rdb.cgi?action=res_entries.

Bouton, M. E. (1994). "Conditioning, Remembering, and Forgetting.*"* *Journal of Experimental Psychology: Animal Behavior Processes,* **20**(3): 219-231.

Cade, B. & O'Hanlon, W. H. (1993).*A Brief Guide to Brief Therapy*. New York W.W. Norton.

Canales, J. J. (2005). "Stimulant-induced adaptations in neostriatal matrix and striosome systems: Transiting from instrumental responding to habitual behavior in drug addiction." *Neurobiology of Learning and Memory,* **83**(2): 93-103.

Centonze, D., Picconi, B., Baunez, C., Borrelli, E., Pisani, A., Bernard, G., & Calabresi, P. (2002). Cocaine and Amphetamine Depress

Striatal GABAergic Synaptic Transmission through D2 Dopamine Receptors. *Neuropsychopharmacology*, 26, 164–175.

Chambers, R. A., J. R. Taylor, et al. (2003). "Developmental Neurocircuitry of Motivation in Adolescence: A Critical Period of Addiction Vulnerability." *The American Journal of Psychiatry,* **160**(6): 1041-1052.

Chambers, R. A., J. R. Taylor, et al. (2003). "Developmental Neurocircuitry of Motivation in Adolescence: A Critical Period of Addiction Vulnerability." *The American Journal of Psychiatry,* **160**(6): 1041-1052.

Chambers, R. A., W. K. Bickel, et al. (2007). "A scale-free systems theory of motivation and addiction." *Neuroscience & Biobehavioral Reviews,* **31**(7): 1017-1045.

Charvet S. R. (1997). *Words That Change Minds: Mastering the Language of Influence.* Dubuque, IA: Kendall Hunt Publishing.

Chesterton, G. K. (1908/1995). *Orthodoxy.* San Francisco: Ignatius Press.

Cloninger C. R. (1987). Neurogenetic adaptive mechanisms in alcoholism. *Science,* 1987(236):410-416.

Colleau, S. M. & Joranson, D. (1998). Fear of addiction: confronting a barrier to cancer pain relief. *Cancer Pain Release.* 11(3). Retrieved on July 5, 2008 from http://whocancerpain.bcg.wisc.edu/old_site/eng/11_3/fear.html

Correctional Service of Canada. (1996). *Substance Abuse Treatment Modalities: Literature Review. Drug and Alcohol Education.* Ottawa: Correctional Service Canada. Retrieved on July 21, 2008 from http://www.csc-scc.gc.ca/text/pblct/litrev/treatmod/lit8e-eng.shtml

Cowles, E. L, Castellano, T. C., & Gransky, L. A. (1995). *"Boot Camp" Drug Treatment and Aftercare Interventions: An Evaluation Review.* Rockville, MD: National Institute of

Justice. Retrieved July 21, 2008 from http://www.ncjrs.gov/txtfiles/btcamp.txt

Craig, A. D. (2002). "How do you feel? Interoception: the sense of the physiological condition of the body." *Nature Reviews: Neuroscience,* **3**(8): 655-666.

Craig, A. D. (2009). "How do you feel--now? The anterior insula and human awareness." *Nature Reviews: Neuroscience,* **10**(1): 59-70.

Czikszentmihalyi, M. (ND). *Flow: The Psychology of Optimal Experience (Steps toward Enhancing the Quality of Life).* Retrieved on July 24, 2008 from http://web.ionsys.com/~remedy/FLOW%20%20.htm

Csikszentmihalyi, M. & Csikszentmihalyi, I. S. (1988). *Optimal Experience: Psychological Studies of Flow in Consciousness.* Cambridge: Cambridge University Press.

Czikszentmihalyi, M. (1991). *Flow: The psychology of optimal experience.* NY: Harper Perennial Editions.

Damasio, A. R. (1999). *The Feeling of What Happens: Body and Emotion in the Making of Consciousness.* New York: Harcourt.

Dass, R. (1971). *Be Here Now: Dr. Richard Alpert, Ph. D., Into Baba Ram Dass.* NY: Crown Publishing Group,

Davidson, R. J. (1993). Parsing Affective Space: Perspectives from Neuropsychology and Psychophysiology. *Neuropsychology,* 7(4), 464-475.

Day, R. C. G. (1985). Students' perceptions of Neurolinguistic Programming strategies (counseling, communication, clients, therapy) (Doctoral Dissertation, Florida State University, 1985). *Dissertation Abstracts International,* 46(4), 1333. Retrieved November 24, 2006 from http://www.nlp.de/cgi-bin/research/nlp-rdb.cgi?action=res_entries.

Deci, E. L., & Ryan, R. M. (2008). Facilitating Optimal Motivation and Psychological Well-Being across Life's Domains. *Canadian Psychology,* 49(1), 14–23.

DeLozier, J., & Grinder, J. (1987). *Turtles All The Way Down: Prerequisites for Personal Genius*. Santa Cruz, CA: Grinder, DeLozier and Associates.

DiClemente, C. (2006, June). "The Stages of Addiction Treatment and Social Work Practice: From Prevention to Aftercare" Keynote Presentations at the National Association of Social Workers Annual Addictions Institute, Fordham University, Lincoln Center, NY, NY

DiClemente, C. C. (1994). If Behaviors Change Can Personality be Far Behind? In T. F. Heatherton & J. L. Weinberger (eds.), *Can Personality Change?* Washington, DC: American Psychological Association.

DiClemente, C. C. (2003). *Addiction and Change: How Addictions Develop and Addicted People Recover*. New York: The Guilford Press.

Diekhof, E. K., P. Falkai, et al. (2008). "Functional neuroimaging of reward processing and decision-making: A review of aberrant motivational and affective processing in addiction and mood disorders." *Brain Research Reviews,* **59**(1): 164-184.

Dilts, R. (1983). *Roots of NLP*. Cupertino, CA: Meta Publications.

Dilts, R. (1993). *Changing Belief Systems with NLP*. Cupertino, CA: Meta Publications.

Dilts, R. (1995). *Strategies of Genius (vol. 3)*. Cupertino CA: Meta Publications.

Dilts, R., & Delozier, J. (2000). *Encyclopedia of Systemic Neuro-Linguistic Programming and NLP New Coding*. Scotts Valley, CA: NLP University Press.

Dilts, R., Delozier, J., Bandler, R. & Grinder, J. (1980). *NLP. Vol.1*. Capitola, CA: Meta Publications.

Doweiko, H. (1996). *Concepts of Chemical Dependency (Third Ed.)*. Pacific Grove, CA: Brooks/Cole.

Driskell, J., Copper, C., & Moran, A. (1994). Does mental practice enhance performance? *Journal of Applied Psychology*, 79(4), 481-492.

Ehrmantraut, J. E., Jr. (1983). A comparison of the therapeutic relationships of counseling students trained in Neurolinguistic Programming vs. students trained on the Carkhuff Model. Doctoral Dissertation, University of Northern Colorado, 1983). *Dissertation Abstracts International*, 44(10), 3191-B. Retrieved November 24, 2006 from http://www.nlp.de/cgi-bin/research/nlp-rdb.cgi?action=res_entries.

Ferster, C. B. & Skinner, B. F. (1953). *Schedules of Reinforcement* New York: Macmillan Free Press.

Freeman, W. J. (1998). The Neurobiology of Multimodal Sensory Integration. *Integrative Physiological & Behavioral Science*, 33(2), 124-129.

Frieden, F. P.: Speaking the client's language: the effects of Neurolinguistic Programming (predicate matching) on verbal and nonverbal behaviors in psychotherapy. A single case design (Doctoral Dissertation, Virginia Commonwealth University, 1981). *Dissertation Abstracts International*, 42(3), 1171-B. Retrieved November 24, 2006 from http://www.nlp.de/cgi-bin/research/nlp-rdb.cgi?action=res_entries.

Gilbert, D. T. & Malone, P. S. (1995). The Correspondence Bias. *Psychological Bulletin*, 117(1), 21-38.

Glasser, William. (1985). *Positive Addiction*. NY: Harper Collins.

Goldstein, R. Z. & Volkow, N. D. (2002). Drug addiction and its underlying neurobiological basis: Neuroimaging evidence for the involvement of the frontal cortex. *American Journal of Psychiatry*, 159(10), 1642-1652.

Goodwyn, E. D. (2012). *The Neurobiology of the Gods: How Brain Physiology Shapes the Recurrent Imagery of Myth and Dream*. New York: Routledge.

Gray, R. M. (1996). *Archetypal Explorations*. London: Routledge.

Gray, R. M. (2001). Addictions and the Self: A Self-Enhancement Model for Drug Treatment in the Criminal Justice System. *The Journal of Social Work Practice in the Addictions*, 2(1). Retrieved April 1, 2006 from http://richardmgray.home.comcast.net

Gray, R. M. (2002). "The Brooklyn Program: Innovative Approaches to Substance Abuse Treatment." *Federal Probation Quarterly vol. 66*(3), 9-16.

Gray, R. M. (2003). The Brooklyn Program: Cognitive applications of the physiological correlates of spiritual experience. *The Dr. Lonnie E. Mitchell National HBCU Substance Abuse Conference*, sponsored by Howard University, on April 2, 2003.

Gray, R. M. (2005). *Thinking About Drugs and Addiction.* Boulder CO: NLP Comprehensive. http://www.nlpco.com/articles/AddictionsGray.html

Gray, R. M. (2008). NLP and Levels of Motivation. *Suppose, the Official CANLP/ACPNL Bilingual Newsletter.* Fall 2008, 20-24.

Gray, R. M. (2011a). *Interviewing and counseling skills: An NLP perspective.* Raleigh, NC: Lulu Press.

Gray, R. M. (2011b). *Transforming Futures: The Brooklyn Program Facilitators Manual Second Edition.* Raleigh, NC: Lulu Press.

Gray, R. M. (2012) Addictions. In Lisa Wake, Richard Gray & Frank Bourke (Eds.), The Clinical Efficacy of NLP: A critical appraisal (95-125). London, Routledge.

Gray, R. M. & Liotta, Richard F. (2012). PTSD: Extinction, Reconsolidation and the Visual-Kinesthetic Dissociation Protocol. *Traumatology. 18*(2), 3-16. DOI 10.1177/1534765611431835.

Green, M. A. (1979). Trust as effected by representational system predicates (Doctoral Dissertation, Ball State University, 1979).

Dissertation Abstracts International, 41(8) 3159-B. Retrieved November 24, 2006 from http://www.nlp.de/cgi-bin/research/nlp-rdb.cgi?action=res_entries.

Hammer, A. L. (1980). Language as a therapeutic tool: the effects on the relationship of listeners responding to speakers by using perceptual predicates (Doctoral Dissertation, Michigan State University, 1980). *Dissertation Abstracts International*, 41 (3), 991-A 149. Retrieved November 24, 2006 from http://www.nlp.de/cgi-bin/research/nlp-rdb.cgi?action=res_entries.

Haule, J. R. (2010*). Jung in the 21st Century* (2 volumes). London: Routledge.

Henderson, M. (2003). Chimp genome helps reveal secrets of man. *The London Times.* September 1, 2005. Retrieved on July 1, 2008 from http://www.timesonline.co.uk/tol/news/world/article561094.ece

Hillman, J. (1977). *Revisioning Psychology*. New York: Harper Colophon.

Hillman, J. (1996). *The Soul's Code: In Search of Character and Calling*. New York: Random House.

Hilts, P. J. (1994, August 2). Is Nicotine Addictive? It Depends On Whose Criteria You Use. *The New York Times.* Retrieved on July 31, 2008 from http://www.marijuanalibrary.org/

Højsted, J., & Sjøgre, P. (2007). Addiction to opioids in chronic pain patients: A literature review. *European Journal of Pain, 11*(5), July 2007, 490-518

Hulleman, C. S., Durik A. M., Schweigert S. A., & Harackiewicz, J. M. (2008).Task Values, Achievement Goals, and Interest: An Integrative Analysis. *Journal of Educational Psychology*, 100(2), 398–416. DOI: 10.1037/0022-0663.100.2.398

Hyman, S. E., Malenka, R. C., & Nestler, E. J. (2006). Neural Mechanisms of Addiction: The Role of Reward-Related

Learning and Memory. *Annual Review of Neuroscience.* 29: 565-598.

IASH & Delozier, J. (2006). An Interview with our Keynote Speaker [Interview Transcript]. Retrieved from IASH 2006 Conference Web site: http://www.nlpiash.org/conference2006/Site/Presentations/DelozierJudith.htm

Koestner, R. (2008). Reaching One's Personal Goals: A Motivational Perspective Focused on Autonomy. *Canadian Psychology,* 49(1), 60-67.

Koob G. F. (1992). Neural mechanisms of drug reinforcement. *Annals of the New York Academy of Sciences,* 654(1), 171-191.

Kringelbach, M L. (2005). The Human Orbitofrontal Cortex: Linking Reward to Hedonic Experience. *Nature Reviews: Neuroscience, 6,* September 2005, 691-702.

Kroes, M. C. W. and G. Fernández (2012). "Dynamic neural systems enable adaptive, flexible memories." Neuroscience & Biobehavioral Reviews **36**(7): 1646-1666.

Lakoff, G. & Johnson, M. (1980). *Metaphors we live by.* Chicago: University of Chicago Press.

Lang, P. J. (1983). Fear Behavior, Fear Imagery and Psychophysiology of Emotion: The Problem of Affective Response Integration. *Journal of Abnormal Psychology,* 92(3), 276-306.

Lang, P. J. (1994). The Varieties of Emotional Experience: A Meditation on James-Lange Theory. *Psychological Review,* 101(2), 211-221.

Laundergan, J. C. (1982). *Easy Does It.* Minneapolis, MN: Hazelden.

Leshner, J. A. (2005). The Essence of Drug Addiction. *The National Institutes of Drug Abuse.* Retrieve on June 28, 2008 from http://www.nida.nih.gov/Published_Articles/Essence.html

Lewis, B. & Pucelik, F. (1990). *Magic of NLP Demystified.* Portland, OR: Metamorphous Press.

Liechti, M. E., Vollenweider, F. X. (2001). Which neuroreceptors mediate the subjective effects of MDMA in humans? A summary of mechanistic studies. *Human Psychopharmacology*, 16: 589-598.

Linden, A & Perutz, K. (1998). *Mindworks: NLP Tools for Building a Better Life*. NY: Berkley Publishing Group.

Lopez-Quintero, C., D. S. Hasin, et al. (2011). "Probability and predictors of remission from life-time nicotine, alcohol, cannabis or cocaine dependence: results from the National Epidemiologic Survey on Alcohol and Related Conditions." *Addiction* **106**(3): 657-669.

Martin, K., & Hall, C. (1995). *Using mental imagery to enhance intrinsic motivation. Journal of Sport & Exercise Psychology*, 17(1), 54-69.

Maslow, A. (1970). *Religions, Values, and Peak Experiences*. New York: The Viking Press.

McClure, S. M., Daw, N. D. & Montague, P. R. (2003). A computational substrate for incentive salience. *Trends in Neuroscience*, 26(8), 423-8.

McKim, W. A. (2003). *Drugs and Behavior: An Introduction to Behavioral Pharmacology (Fifth Ed.)*. Upper Saddle River, NJ: Prentice Hall.

Medina J. L, & Diamond S. (1977). Drug dependency in patients with chronic headaches. *Headache,* 17: 12-14.

Miller, G.A., Galanter, E., & Pribram, K.H. (1960). *Plans and the Structure of Behavior*. New York: Holt, Rinehart & Winston.

Miller, S. D. & Berg, I., K. (1995). *The Miracle Method: A Radically New Approach to Problem Drinking.* NY: Norton.

Miller, W. R. (1995). *Motivational Enhancement Therapy with Drug Abusers* Albuquerque, New Mexico Center on Alcoholism, Substance Abuse, and Addictions (CASAA). Retrieved on July 28, 2008 from http://www.motivationalinter

Miller, W. R. (2004).The Phenomenon of Quantum Change, *Journal of Clinical Psychology: In Session*, 60(5), 453–460

Miller, W. R., Zweben, A., DiClemente, C. C. & Rychtarik, R. G. (1994). *Motivational enhancement therapy manual: A clinical research guide for therapists treating individuals with alcohol abuse and dependence.* Project MATCH Monograph Series, Vol. 2. DHHS Publication No. 94-3723. Rockville MD: NIAAA.

Miller, W. R., & C'de Baca, J. (1994). Quantum change: Toward a psychology of transformation. In T. Heatherton & J. Weinberger (Eds.), *Can personality change?* (pp. 253–280). Washington, DC: American Psychological Association.

Montague, P. R., Hyman, S. E., & Cohen, J. D. (2004). Computational roles for dopamine in behavioral control. *Nature*, 431:760–67.

Morris, R. G. M. (2006). Elements of a neurobiological theory of hippocampal function: The role of synaptic plasticity, synaptic tagging and schemas. *European Journal of Neuroscience*, 23(11), 2829-2846.

Morse, R. M. & Flavin, D. K. (1992). The definition of alcoholism. The Joint Committee of the National Council on Alcoholism and Drug Dependence and the American Society of Addiction Medicine to Study the Definition and Criteria for the Diagnosis of Alcoholism. *JAMA*, 268(8), August 26, 1992, 1012-4.V

Mucha, R. F., van der Kooy, D., O'Shaighnessy, M., & Bucenicks, P. (1982). Drug reinforcement studies by the use of place conditioning in rat. *Brain Research, 243,* 91-105.

Nadel, L., A. Hupbach, et al. "Memory formation, consolidation and transformation." *Neuroscience & Biobehavioral Reviews,* 36(7):1640-5

Nader, K., Bechara, A. & van der Kooy, D. (1997). Neurobiological constraints on behavioral models of motivation. *Annual Review of Psychology.* Palo Alto, CA (48), 85-114.

Naqvi, N., & Bechara, A. (2009). The hidden island of addiction: the insula. *Trends in Neurosciences, 32*(1), 56-67. doi: http://dx.doi.org/10.1016/j.tins.2008.09.009

Naqvi, N., Rudrauf, D., Damasio, H., & Bechara, A. (2007). Damage to the Insula Disrupts Addiction to Cigarette Smoking. *Science, 315*(5811), 531-534. doi: 10.1126/science.1135926

National Institutes of Drug Abuse (NIDA). (2002). *Stress and Substance Abuse: A Special Report. National Institute on Drug Abuse (NIDA)*. http://www.drugabuse.gov/stressanddrugabuse.html. Tuesday, February 26, 2002

New York University (NYU). (2008, June 30). Using Mental Strategies Can Alter The Brain's Reward Circuitry. *Science Daily*. Retrieved June 30, 2008, from http://www.sciencedaily.com/releases/2008/06/080629130753.htm

Notz, W. W. (1975). Work Motivation and the Negative Effects of Extrinsic Rewards. *American Psychologist,* (September 1975), 884-891

O'Connor, J., & Seymour, J. (1990). *Introducing NLP.* London: Element.

Olds, J., & Milner, P. (1954). Positive reinforcement produced by electrical stimulation of septal area and other regions of rat brain. *Journal of Comparative Physiological Psychology,* 47:419-27.

Overdurf, J. (2006, April). *You Never Know How Far a Change Will Go ...Beyond Goals.* Pre-Conference workshop conducted at the 19th Annual Convention of the Canadian Association of NLP. Retrieved on April 15, 2008 from http://johnoverdurf.typepad.com/canlp/files/canlpmanual.pdf

Palubeckas, A. J. (1981). Rapport in the therapeutic relationship and its relationship to pacing (Doctoral Dissertation, Boston

University School of Education, 1981). *Dissertation Abstracts International*, 42(6), 2543-B 2544-B. Retrieved November 24, 2006 from http://www.nlp.de/cgi-bin/research/nlp-rdb.cgi?action=res_entries.

Pantin, H. M. (1982). The relationship between subjects' predominant sensory predicate use, their preferred representational system and self-reported attitudes towards similar versus different therapist-patient dyads (Doctoral Dissertation University of Miami, 1982). *Dissertation Abstracts International*, 43(7), 2350-B. Retrieved November 24, 2006 fromhttp://www.nlp.de/cgi-bin/research/nlp-rdb.cgi?action=res_entries.

Peck, M. S. (1998). *The Road Less Traveled: A New Psychology of Love, Traditional Values and Spiritual Growth (Third Ed.)*. NY: Simon & Schuster.

Peele, S. (1987). Why Do Controlled-Drinking Outcomes Vary by Investigator, by Country and by Era? Cultural Conceptions of Relapse and Remission in Alcoholism. *Drug and Alcohol Dependence*, 20:173-201.

Peele, S. (1989). *Diseasing of America: Addiction treatment out of control*. Lexington, MA: Lexington Books.

Peele, S., Brodsky, A. & Arnold, M. (1992). *The Truth About Recovery and Addiction*. New York: Simon & Schuster.

Perry S. & Heidrich G. (1882). Management of pain during debridement: a survey of U.S. burn units. *Pain,* 13:267-280.

Pesoa, L. (2008). On the relationship between emotion and cognition. *Nature Neuroscience*, 9:148-158.

Piaget, J. (1970). *Genetic Epistemology* (Eleanor Duckworth, Trans.). New York: Columbia University Press

Portenoy R. K., & Foley K. M. (1986). Chronic use of opioid analgesics in nonmalignant pain: Report of 38 cases. *Pain,* 25: 171-186.

Porter J, & Jick H. (1980). Addiction rare in patients treated with narcotics. *N Engl J Med.,* 302: 123.

Price, R. K., Risk, N. K. et al. (2001). "Remission from drug abuse over a 25-year period: patterns of remission and treatment use." Am J Public Health, **91**(7): 1107-1113.

Prochaska, J. O. (1979). *Systems of psychotherapy: A transtheoretical analysis*. Homewood, IL: Dorsey Press.

Prochaska, J. O. (1994). Strong and Weak Principles for Progressing From Precontemplation to Action on the Basis of Twelve Problem Behaviors. *Health Psychology*, 13(1): 47-51.

Prochaska, J. O., Norcross, J. C., & DiClemente, C. C. (1994). *Changing for Good*. New York: William Morrow.

Prochaska, J. O., Di Clementi, C. C. & Norcross, J. C. (1992). In search of How People Change: Application to Addictive Behaviors. *American Psychologist*, 5(9): 1102-1114.

Raine, A. (1993). The Psychopathology of Crime. San Diego: Academic Press.

Raine, A., et al. (1997). "Brain abnormalities in murderers indicated by positron emission tomography." Biol. Psychiatry 42(6): 495-508.

Ramskogler, K., H. Walter, et al. (2001). "Subgroups of Alcohol Dependence and their Specific Therapeutic Management: A Review and Introduction to the Lesch- Typology." *International Addiction,* **May 2001**(2). Retrieved from http://www.isamweb.com/pages/pdfs/e-book%20Issue%202/Lesch.pdf

Rat Genome Sequencing Project Consortium. (2004). Genome sequence of the Brown Norway rat yields insights into mammalian evolution. *Nature, 428*, 493-521 (1 April 2004).

Redish, A. D., Jensen, S. et al. (2008). "A unified framework for addiction: Vulnerabilities in the decision process." *Behavioral and Brain Sciences,* **31**(04): 415-437.

Rescorla, R. A. (1988). "Pavlovian conditioning: It's not what you think it is." *American Psychologist,* **43**(3): 151-160.

Robbins, T. W., Gillan, C. M. et al. (2012). "Neurocognitive endophenotypes of impulsivity and compulsivity: towards dimensional psychiatry." *Trends in Cognitive Sciences,* **16**(1): 81-91.

Robins, L. N. (1973). *The Vietnam Drug User Returns.* Washington D.C.: U.S. Government Printing Office.

Robins, L. N., Davis, D. H. & Nurco, D. N. (1974). How Permanent Was Vietnam Drug Addiction? *American Journal of Public Health. 64(12 Suppl): 38–43.*

Robins, L. N., Helzer, J. E. & Davis, D. H. (1975). Narcotic use in Southeast Asia and afterward. *Archives of General Psychiatry.* 32(8), 955-61.

Robinson, T. E. & Berridge, K. C. (2001). "Incentive-sensitization and Addiction." *Addiction*, 96(1), 103–114.

Robinson, T. E. (2004). "Addicted Rats." *Science*, 305(5686), 951-953.
DOI: 10.1126/science.1102496.

Robinson, T. E. & Berridge, K. C. (2003). Addiction. *Annual Review of Psychology,* **54**, 25-53. DOI: 10.1146/annurev.psych.54.101601.145237

Rossi, E. L. (1986). *The Psychobiology of Mind-Body Healing.* NY: W.W. Norton.

Ruden, R. (1997). *The Craving Brain.* New York: Harper Collins.

Sandhu, D. S.; Reeves, T. G.; Portes, P. R. (1993). Cross-cultural counseling and neurolinguistic mirroring with Native American adolescents. *Journal of Multicultural Counseling and Development,* 21(2) 106-118. Retrieved on November 25, 2006 from PsychArticles.

Schacter, D. L. & Addis, D. R. (2007a). Constructive memory. The ghosts of past and future. *Nature*, 445(4), 27. Doi: 10.1038/445027a

Schacter, D.L. & Addis, D.R. (2007b). The cognitive neuroscience of constructive memory: Remembering the past and imagining the future. *Philosophical Transactions of the Royal Society (B), 362,* 773-786.

Schaeffer Library of Drug Policy (N.D.). *Addiction Criteria.* Retrieved June 27, 2008 from http://www.druglibrary.org/Schaffer/library/addcrit.htm

Schmedlen, G W. (1981). The impact of sensory modality matching on the establishment of rapport in psychotherapy (Doctoral Dissertation, Kent State University, 1981). *Dissertation Abstracts International,* 42(5), 2080-B. Retrieved November 24, 2006 from http://www.nlp.de/cgi-bin/research/nlp-rdb.cgi?action=res_entries.

Schug. S. A., Zech, D., Grond, S., Jung, H., Meuser, T. & Stobbe, B. (1992). A long-term survey of morphine in cancer pain patients. *J Pain Symptom Manage,* 1992(7), 259-66.

Schultz, W., Dayan, P. & Montague, P. R. (1997). A neural substrate of prediction and reward. *Science,* 275, 1593-1599.

Shattuck, D. K. (1994). Mindfulness and metaphor in relapse prevention: an interview with G Alan Marlatt. *Journal of the American Dietetic Association,* 94(8):846-8.

Shell, D. F., & Husman, J. (2008). Control, Motivation, Affect, and Strategic Self-Regulation in the College Classroom: A Multidimensional Phenomenon. *Journal of Educational Psychology, 100(2):443-459*

Shobin, M Z. (1980). An investigation of the effects of verbal pacing on initial therapeutic rapport (Doctoral Dissertation, Boston University School of Education, 1980). *Dissertation Abstracts International,* 41(5). Retrieved November 24, 2006 from http://www.nlp.de/cgi-bin/research/nlp-rdb.cgi?action=res_entries.

Siegel, S. (1984). Pavlovian conditioning and heroin overdose: Reports by overdose victims. *Bulletin of the Psychonomic Society,* 22, 428-430.

Siegel, S., Hinson, R.E., Krank, M.D. & McCully, J. (1982). Heroin "overdose" death: contribution of drug-associated environmental cues. *Science,* 216, 436-437.

Stimmel, B. (2002). *Alcoholism, Drug Addiction and the Road to Recovery: Life on the Edge.* New York: The Haworth Medical Press.

Thomason, D. D. (1984). Neurolinguistic Programming: an aid to increase counselor expertness (Doctoral Dissertation, Biola University, 1984). *Dissertation Abstracts International*, 44(9), 2909-B. Retrieved November 24, 2006 from http://www.nlp.de/cgi-bin/research/nlp-rdb.cgi?action=res_entries.

Tobler, P. N.; Fiorillo, Christopher, D. & Schultz, W. (2005). Adaptive Coding of Reward Value by Dopamine Neurons. *Science*, 307, 1642-1645.

Treasure, J. (2004). Motivational interviewing. Advances in Psychiatric Treatment), 10: 331–337.

Tse, D., Langston, R. F., Bethus, I., Wood, E. R., Witter, M. P., & Morris, R. G. M. (2008). Does assimilation into schemas involve systems or cellular consolidation? It's not just time. *Neurobiology of Learning and Memory*, 89(4), 361-365. doi: 10.1016/j.nlm.2007.09.007

United Nations (UN) (2003). *Investing in Drug Abuse Treatment: A Discussion Paper for Policy Makers.* New York: United Nations. Retrieved July 21, 2008 from http://www.unodc.org/docs/treatment/Investing_E.pdf

Waelti, P., Dickenson, A. & Schults, W. (2001). Dopamine responses comply with basic assumptions of formal learning theory. *Nature*, Vol. 412, 43-48.

Waldorf, D. & Biernacki, P. (ND). Natural Recovery from Heroin Addiction: A Review of the Incidence Literature. Retrieved on June 25, 2007 from http://www.drugtext.org/library/articles/narehead.htm

Waldorf, D. (1983). ""Natural recovery from opiate addiction: Some social-psychological processes of untreated recovery." *Journal of Drug Issues,* **83**(2): 237-280.

Waldorf, D. and P. Biernacki (1979). "Natural recovery from heroin addiction: A review of the incidence literature." *Journal of Drug Issues* (Spring): 281-288.

Watzlawick, P. (1978). *The Language of Change: Elements of Therapeutic Communication.* New York: W. W. Norton.

Watzlawick, P., Weakland, J. & Fisch, R. (1974). *Change: Principles of Problem Formation and Problem Resolution.* New York: W. W. Norton.

Weaver, S. C. & Tennant F. S. (1973). Effectiveness of Drug Education Programs for Secondary School Students. *American Journal of Psychiatry*, 130:812-814, July 1973.

Wegner, D. M., Schneider, D. J., Carter, S., & White, T. (1987). Paradoxical effects of thought suppression. *Journal of Personality and Social Psychology*, 53, 5-13.

Williams, L. & Bargh, J. (2008). Experiencing physical warmth promotes interpersonal warmth. *Science, 322*, 606-607.

Wilson, C., (2002). *Beyond the Occult: Twenty Years Research into the Paranormal.* NY: Carroll & Graf

Wohldmann, E., Healy, A., & Bourne, L. (2007). *Pushing the limits of imagination: Mental practice for learning sequences. Journal of Experimental Psychology: Learning, Memory, and Cognition*, 33(1), 254-261. doi:10.1037/0278-7393.33.1.254.

Young, F. L. (September 6, 1988). Opinion and Recommended Ruling, Findings of Fact, Conclusions of Law and Decision of Administrative Law Judge. *In The Matter Of Docket No. 86-22 Marijuana Rescheduling Petition,* Section VIII. Accepted Safety For Use Under Medical Supervision: findings of fact, paragraphs 8 & 9.

Zenz, M., Strumpf, M. & Tryba, M. (1992). Long-term oral opioid therapy in patients with chronic nonmalignant pain. *J Pain Symptom Manage,* 7: 69-77.

Zoja, L. (1990). *Drugs, Addiction & Initiation: The Modern Search for Ritual*. Gloucester, MA: Sigo.